Solutions Manual to Accompany

Fundamentals of Quality Control and Improvement

Solutions Manual to Accompany

Fundamentals of Quality Control and Improvement

Third Edition

AMITAVA MITRA

Auburn University
College of Business
Auburn, Alabama

WILEY

A JOHN WILEY & SONS, INC., PUBLICATION

Published by John Wiley & Sons, Inc., Hoboken, New Jersey.
Published simultaneously in Canada.

For general information on our other products and services or for technical support, please contact our Customer Care Department within the United States at (800) 762-2974, outside the United States at (317) 572-3993 or fax (317) 572-4002.

Wiley also publishes its books in a variety of electronic formats. Some content that appears in print may not be available in electronic format. For information about Wiley products, visit our web site at www.wiley.com.

Library of Congress Cataloging-in-Publication Data:

Mitra, Amitava
   Solutions manual to accompany Fundamentals of Quality Control and Improvement—3rd edition

   ISBN 978-0-470-25697-8 (paper)

10 9 8 7 6 5 4 3 2 1

CONTENTS

PREFACE

This solutions manual is designed to accompany the text, "Fundamentals of Quality Control and Improvement." To assist the student and the instructor in the teaching of the material, this manual includes solutions to the end-of-the chapter problems. The answers to the discussions questions are included too. Detailed explanation on the discussion questions may be found in the text and references. Associated figures and graphs on solutions to the problems are kept to a minimal. Most of the computations may be conducted using the Minitab software.

CHAPTER 1

INTRODUCTION TO QUALITY CONTROL AND THE TOTAL QUALITY SYSTEM

1-1.   a)   Call center that sells computers – possible definitions of quality that involve different variables/attributes could be as follows:

i)   Time to process customer order for computers – Time measured in hours.
ii)   Total turn over time (starting with customer placement of order to customer receipt of computer) – Time measured in hours.
iii)   Proportion of delivered orders that do not match customer requirements exactly.
iv)   Proportion of orders that are fulfilled later than promised date.

Integration of the various measures to one measure is not easily attainable. Individual measures, as proposed, should not be difficult to measure.

b)   Emergency services for a city or municipality:

i)   Time to respond to an emergency – Time measured in minutes.
ii)   Time to process an emergency call – Measured in minutes and seconds.

Proposed measures readily obtainable.

c)   Company making semiconductor chips:

i)   Total manufacturing costs/10,000 chips.
ii)   Parts per million of defective chips.
iii)   Equipment and overhead costs/10,000 chips.

Measure iii) can be integrated into measure i). Measure ii) will influence manufacturing costs per conforming product. All of the measures should be easily obtainable.

d)   A hospital: Variety of measures exist based on patient satisfaction, effectiveness of services, efficiency of operations, rate of return to investors, and employee/staff/nurse/physician satisfaction.

i)   Proportion of in-patients satisfied with services.
ii)   Length of stay of patients, by specified diagnosis related groups – Measured in days.
iii)   Turn around time for laboratory tests, by type of test – Measured in hours/minutes.
iv)   Annual or quarterly rate or return.

Most of the measures can be readily obtained. It may be difficult to integrate all such measures. However, some of these measures, such as annual rate of return, may serve as an integrated measure.

e)   Deliver mail/packages on a rapid basis:

2

i) Total turn around time (from taking order to delivery) for packages – Measured in hours.
ii) Processing time of orders – Measured in minutes.
iii) Proportion of packages not delivered within promised time.
iv) Proportion of packages delivered to wrong address/person.

All of these measures should be easily obtainable. Measure ii) obviously is part of measure i). Measure i) may also influence measure iii). Measures iii) and iv) may involve causal analysis to identify reasons for errors or long delivery times. Measures i) and ii) could be analyzed for improving efficiency of the process.

f) A department store – Several forms of measures exist based on customer satisfaction, employee satisfaction, and rate of return to investors.

i) Proportion of customers satisfied with the store services.
ii) Time taken to service individual customers – Measured in minutes.
iii) Waiting time of customers before being serviced – Measured in minutes.
iv) Proportion of staff turnover.
v) Annual or quarterly rate of return to investors.

Majority of the proposed measures can be obtained with reasonable ease. Some serve as an integrated measure, for example, annual rate of return to investors.

g) A bank – Several forms of measure exist based on customer satisfaction, employee satisfaction, or rate of return to investors.

i) Proportion of customers satisfied with the bank services.
ii) Total time taken to serve the bank customer – Measured in minutes.
iii) Waiting time of customers before being serviced – Measured in minutes.
iv) Proportion of staff turnover.
v) Annual rate of return to investors.

Majority of the measures can be obtained with reasonable ease. Some of these serve as an integrated measure, for example, annual rate of return to investors.

h) A hydro-electric power plant – Several operational, effectiveness, and financial measures exist:

i) Cost per kilowatt-hour of electricity produced – Measured in dollars and cents.
ii) Total kilowatt-hours produced monthly – Influenced by demand.
iii) Proportion of total customer demand met by particular plant.
iv) Annual rate of return to investors.

Most of these measures can be obtained with reasonable ease. Some of these serve as an integrated measure, for example, annual rate of return to investors.

1-2.   Quality of design – Ensure total service time to the customer or alternatively waiting time to the customer is minimized. Ensure a variety of services demanded by customers are provided. For example, such may include guidelines on investment, home mortgage loans, home improvement loans, automobile loans, financial management services for the elderly, availability of several locations that are of proximity to customers, etc. Quality of conformance should address the means to achieve the variety of features that are discussed in the design stage. Quality of performance will finally address and measure how the bank does in meeting the desired goals when it is operational. Some measures in this performance phase could be:

i)   Percentage of customers satisfied with all services.
ii)   Percentage of customers satisfied with financial management services.
iii)   Dollar volume of loans processed per month.
iv)   Time to respond to customer inquiry – Measured in minutes.

Basic needs in this context could consist of the following: Offer a variety of checking/savings accounts, safe deposit boxes, several ATM locations in convenient places easily accessible to customers. Performance needs could be measured by time to respond to customer inquiry, waiting time of customers, time to process loan application, etc. Excitement needs could consist of special services for customers over the age of 50 years, investment planning assistance, attractive savings/investment promotions that become the benchmark in the industry, remote service locations in buildings with major employers/entertainment/shopping, cash advance with no interest for very short term periods, such as a week, etc.

1-3.   The travel agency should consider improving on the various performance needs, relative to the existing competitors, and possibly providing some of the excitement needs. Obviously, basic needs are assumed to be provided by the travel agency. Some performance needs could be measured by the following: Turn around time per customer, i.e., the total time to provide the customer with the requested service; cost of providing the service; time to respond to a telephone call from a customer; accuracy in fulfilling customer requirements. Some excitement needs could be measured by the following: Meeting with the customer in a convenient location (i.e., place of employment or home); delivery of travel documents to home personally; updating customer with additional promotional/savings features on travel packages even after packet has been delivered, etc.

Impact on the various costs will be as follows: For basic and performance needs, process costs will likely increase. To improve response time, more agents or more convenient locations might be necessary. To reduce external failure costs, which is equivalent to improving customer satisfaction with the provided services, either additional services will have to be provided through an increase in process investment costs (personnel, facilities, etc.) or the efficiency of services will have to be improved. This will also necessitate added process costs. Internal failure costs (detecting inaccurate travel documents before delivery to customers) can be reduced through additional training of existing staff, so that fewer errors are made or through automated error detection, where feasible, through audit of certain documents. Appraisal costs, thus, could go up initially.

1-4. In the hospitality industry, as in others, special causes could be detected by quality control procedures. On the other hand, common causes may be addressed through quality improvement procedures. Typically quality control methods involve the use of control charts, through selected variables or attributes. Quality improvement methods could involve Pareto analysis, flow chart analysis, cause-and-effect analysis, failure modes and effect analysis, and quality function deployment analysis of the process through cross-functional teams.

Some special causes are delay or long waiting time for customer to check-in due to admission staff not being trained in certain tasks, long time to respond to room requests to deliver food or other items, and conference or banquet rooms being unable due to lack of adequate scheduling processes. Some common causes, that are inherent to the process, whose remediation will require making corresponding process changes could be: Delay in responding to customer requests due to shortage of available staff on duty, inability to provide a reservation due to lack of availability of rooms, inability to meet customer expectations to provide information on tourist attractions in the neighborhood due to lack of training of concierge staff, and so forth.

1-5. For the OEM considering an improvement in its order processing system with its tier-one suppliers, some measures of quality are as follows: Time to process order by the supplier; lead time required by the supplier to deliver component or sub-assembly; proportion of time order is delivered on time; proportion of time order is error-free; and parts-per-million (ppm) of components or sub-assemblies that are nonconforming. Some special causes, in this context, could be: Increased time to process order due to malfunction in order approval process or downtime of computers; increased lead time due to longer lead time in delivery of components by tier-two supplier; increased downtime of certain machine/equipment in tier-one supplier; or wrong setting or equipment used causing increased nonconformance rate. Some common causes, in this environment, could be: Increased time to process order by supplier due to lack of adequate staff/ equipment; increased lead time to deliver sub-assembly due to lack of capacity in tier-one plant; or increased parts-per-million of nonconforming product due to poor quality in shipment of components from tier-two supplier.

1-6. For an inter-modal company, some examples of prevention costs are: Design of an effective tracking system that can locate the specific location of each container at any instant of time; design of a system that flags items once actual schedules deviate from expected schedules based on due dates; and projecting labor requirements based on varying demand. Examples of appraisal costs are: Determination of loading/unloading time from one mode (say, ship) to another (say, train); determination of transportation time of a container from one location to another; and determining percentage of shipments that are late. Examples of internal failure costs are: Rectification of a delayed movement between two stations in order to meet deadline on meeting the delivery time at final destination – such could be accomplished through additional operators and equipment (say, trucks). Examples of external failure costs are those due to not meeting delivery time of goods to final destination and thereby incurring a penalty (per contract). Other examples are loss of market share (or customers) due to goods being damaged on delivery at final destination and thereby having to pay a premium for these goods, incurring a loss in revenue. Customer dissatisfaction due to delivery beyond promised date or goods being damaged could lead to non-renewal of future orders or switching by the customer to a competitor. Such lost orders would be examples of external failure costs.

1-7.   With the advent of a quality improvement program, typically prevention and appraisal costs will increase during the initial period. Usually, as quality improves with time, appraisal costs should decrease. As the impact of quality improvement activities becomes a reality, it will cause a reduction in internal failure and external failure costs, with time. In the long term, we would expect the total quality costs to decrease. The increase in the prevention and appraisal costs should, hopefully, be more than offset by the reduction in internal failure and external failure costs.

1-8.   a)   Vendor selection – Prevention.

        b)   Administrative salaries – Usually staff salaries are in the category of prevention. If there are administrative staffs dedicated to appraisal activities, such as processing of paperwork for audit activities, such salaries could be listed in the appraisal category.

        c)   Downgraded product – Internal failure.

        d)   Setup for inspection – Appraisal.

        e)   Supplier control – Appraisal.

        f)   External certification – Prevention.

        g)   Gage calibration – Appraisal.

        h)   Process audit – Prevention.

1-9.   Labor base index – This index could measure quality costs per direct-labor hour or direct-labor dollar and is commonly used at the line management level. For products or services that are quite labor intensive (for example, transportation by truck, processing of income-tax forms), this could be an appropriate measure. In case there are major changes in wage-rates or inflation, quality costs per labor dollar would be monitored. The cost base index includes quality costs per dollar of manufacturing costs, where manufacturing costs include direct-labor, material, and overhead costs. Thus, in a laboratory in a hospital, processing of X-rays incur technical personnel time and major equipment costs. So, processing or internal failure costs, in such a setting, could be monitored through such an index. It could be used by the hospital administration coordinator. The sales base index, that measures quality costs per sales dollar, is used by senior management, for example the CEO or the COO of an organization. Hence, for a senior executive in the automobile industry, a measure of performance to be monitored could be quality costs as a percentage of sales. Quality costs, in this instance, should capture internal failure and external failure costs (due to customer dissatisfaction and warranty claims).

1-10. It is quite possible to increase productivity, reduce costs, and improve market share at the same time. Through quality improvement activities, one could eliminate operations and thereby reduce production costs as well as production time. When production time is reduced, it leads to improved efficiency, which in effect increases capacity. Thus productivity is improved and costs are reduced. Additionally, with an improvement in quality, customer satisfaction is improved, which leads to an increase in market share through an expanded customer base.

1-11. External failure costs are influenced by the degree of customer satisfaction with the product or service offered. Such influence is impacted not only by the level of operation of the selected organization, but also its competitors, and the dynamic nature of customer preferences. Hence, even if a company maintains its current level of efficiency, if it does not address the changing needs of the customer, external failure costs may go up since the company does not keep up with the dynamic customer needs. Furthermore, if the company begins to trail more and more relative to its competitors, even though it maintains its current level of first-pass quality, customer satisfaction will decrease, leading to increased external failure costs.

1-12. The impact of a technological breakthrough is to shift the location of the total prevention and appraisal cost function, leading to a decrease in such total costs for the same level of quality. This cost function usually increases in a non linear fashion with the quality level q. Additionally, the slope of the function will also reduce at any given level of quality with a technological breakthrough. Such breakthroughs may eventually cause a change in the slope of the prevention and appraisal cost function from concave to convex in nature, beyond a certain level of quality. As indicated in a previous question, the failure cost function (internal and external failures) is influenced not only by the company, but also by its competitors and customer preferences. Assuming that, through the breakthroughs, the company is in a better position to meet customer needs and has improved its relative position with respect to its competitors and has approached (or become) the benchmark in its industry, the failure cost function will drop, for each quality level, and its slope may also decrease, at each point, relative to its former level. Such changes may lead to a target level of nonconformance to be zero.

1-13. Note that the goal of *ISO 14000* is to promote a social responsibility towards sustainability and the use of natural resources. It emphasizes a worldwide focus on environmental management. Thus, as natural resources become scarce, for example, the availability of fossil fuel, the adoption of such standards on a world-wide basis will create an environment for future operations in all manufacturing situations. Adoption of such standards will impact corporate culture and management ethics.

1-14. The monitoring of supply chain quality will be influenced by the type of configuration of the supply chain – dedicated supply chain or a tiered supply chain. In a dedicated supply chain, the supply chain consists of certain suppliers who provide the OEM with components or sub-assemblies. The OEM provides the finished product to certain distributors, that are responsible for meeting customer demand. The same distributor could serve more then one OEM, as also the same supplier. In this type of supply chain structure, different supply chains compete against each other. Thus, for a given supply chain, the quality of the supply chain could be monitored through the following functions: On-time shipment of components or sub-assemblies by suppliers to the OEM, maintaining short lead time by suppliers, maintaining or improving parts-per-million of nonconforming product by suppliers and maintaining or improving unit cost by suppliers. For the OEM, similar criteria could be: Assembly time per product unit, total lead time at the product level, total cost per unit at the product level, and nonconformance rate at the product level.

When the type of supply chain structure is a tiered type, several suppliers at a higher level (say tier 2) provide parts or components to the next level (say tier 1) where sub-assemblies are produced. Next, the various sub-assemblies are collected by an infomediary. The various OEMs draw from this common infomediary to make their final product. As in the other case, the finished product is provided by the OEM to various distributors. However, in this situation, each distributor serves only one OEM. Thus, in addition to some of the measures discussed in the previous context, here are some additional process measures in this context: For a given OEM, the effectiveness of its distributors as measured by proportion of customers satisfied, proportion of market share captured by a distributor, and total proportion conforming at the product level produced by the OEM. For the suppliers that feed their components and sub-assemblies to an infomediary, the quality measures adopted would apply to each of the OEMs, since the OEMs draw from this common infomediary.

1-15. a) Using the data provided, Table 1-1 shows the calculations for overhead rate using the unit-based allocation method.

Using the calculated overhead rate of 77.263%, the cost per unit of each product using the unit-based costing method is shown in Table 1-2.

b) Calculations of the cost per unit of each product using the activity-based costing method are shown in Table 1-3.

Product-unit related costs: Setup and testing: $1.1 million ÷ 63000 = $17.46/unit.
Product-line related costs: CPU C1: $0.5 million ÷ 10,000 = $50/unit.
CPU C2: $1.5 million ÷ 15,000 = $100/unit.
Monitor M1: $0.8 million ÷ 18,000 = $44.44/unit.
Monitor M2: $2.5 million ÷ 20,000 = $125/unit.
Production-sustaining costs: $0.6 million ÷ $9.06 million = 0.066225 = 6.6225% of direct labor costs.

TABLE 1-1. Overhead Rate Using Unit-Based Allocation

| | CPU | | Monitor | | |
	C1	C2	M1	M2	Total
Annual Volume	10,000	15,000	18,000	20,000	
Direct labor $/unit	80	140	120	200	
Total direct labor cost (million $)	0.80	2.10	2.16	4.00	9.06
Total overhead (million $)					7.0
Overhead rate					77.263%

c) As can be observed from a comparison of the unit costs from Table 1-2 and 1-3, here are some inferences. Complex products will typically require higher product-line costs. Thus, the activity-based costing method, that makes proportional allocations, will be a better representation compared to unit-based costing method. Note that among CPUs, model C2 is more complex relative to C1. The unit-based method estimates the unit cost for C2 as $408.17, which is quite less relative to $426.73, as estimated by the activity-based method. The unit-based method, in this situation, will under-cost complex products. A similar result is observed for monitor M2, the more complex of the two monitors. Here, however, the difference between the unit costs in using the unit-based method ($574.53) and the activity-based method ($575.71) is not as significant as that for the CPUs.

1-16. a) Since setup and testing costs are different for CPUs and monitors; we calculate these for each product type.
Product-unit related costs: Setup and testing:
    CPU: $0.4 million $\div$ 25,000 = $16/unit
    Monitor: $0.7 million $\div$ 38,000 = $18.42/unit
Table 1-4 shows the unit costs using the activity-based costing method.

b) In comparing the results of this problem with those in Problem 1-15, we note that unit costs, using the activity-based costing method, have increased for the monitors and have decreased for the CPUs. Observe that the setup and testing costs are higher for monitors than for CPUs, which could have caused this to happen.

TABLE 1-2. Cost Using Unit-Based Allocation

| Cost Components | CPU | | Monitor | |
	C1	C2	M1	M2
Director labor ($)	80	140	120	200
Direct material ($)	60	100	80	120
Assembly ($)	40	60	60	100
Overhead (77.263% of direct labor)	61.81	108.17	92.72	154.53
Total unit cost ($)	241.81	408.17	352.72	574.53

TABLE 1-3. Unit Cost Using Activity-Based Allocation

Cost Components	CPU		Monitor	
	C1	C2	M1	M2
Director labor ($)	80	140	120	200
Direct material ($)	60	100	80	120
Assembly ($)	40	60	60	100
Overhead				
Product unit	17.46	17.46	17.46	17.46
Product-line related	50	100	44.44	125
Production-sustaining (6.6225%)	5.30	9.27	7.95	13.25
Total unit cost ($)	252.76	426.73	329.85	575.71

1-17. a)   We are assuming that the product-line cost ($2.5 million) associated with M2 no longer exists. Further, a corresponding reduction in the total setup and testing costs occur due to not producing M2. With the product-unit setup and testing cost remaining at $17.46/unit, since a total of 43000 units is produced, the total setup and testing cost is $750,780. We are assuming that the other company costs remains at $0.6 million annually.

        The total direct labor costs are now $5.06 million. Hence, the overhead rate for production-sustaining costs is $0.6 million ÷ $5.06 million = 0.11858 = 11.858% of direct labor costs. Table 1-5 shows the unit cost of the products using activity-based costing method.

    b)   By not producing monitor M2, the annual overhead cost reduction to the company = Reduction in setup and testing + Reduction in product-line M2 cost.

$$\text{Reduction} = \$(1.1 - 0.75078) \text{ million} + \$2.5 \text{ million}$$
$$= \$2.84922 \text{ million.}$$

TABLE 1-4. Unit Cost Using Activity-Based Costing Method

Cost Components	CPU		Monitor	
	C1	C2	M1	M2
Director labor ($)	80	140	120	200
Direct material ($)	60	100	80	120
Assembly ($)	40	60	60	100
Overhead				
Product unit	16	16	18.42	18.42
Product-line related	50	100	44.44	125
Production-sustaining (6.6225% of direct labor)	5.30	9.27	7.95	13.25
Total unit cost ($)	251.30	425.27	330.81	576.67

TABLE 1-5. Unit Cost Using Activity-Based Costing Method

Cost Components	CPU		Monitor
	C1	C2	M1
Director labor ($)	80	140	120
Direct material ($)	60	100	80
Assembly ($)	40	60	60
Overhead			
Product unit	17.46	17.46	17.46
Product-line related	50	100	44.44
Production-sustaining (11.858% of direct labor)	9.49	16.60	14.23
Total unit cost ($)	256.95	434.06	336.13

If the company chooses to outsource M2, the amount to be annually paid to the supplier = 20000 x 480 = $9.6 million. Hence, net outflow annually = $6.75078 million.

If the company produces monitor M2 (using the previous data), the added cost relative to not producing it:

Added Cost = Direct costs + added overhead
= $420 x 20,000 + ($0.34922 + $2.5) million
= $11.24922 million.

So, the decision is to outsource monitor M2.

1-18. a) Overhead costs (per 1000 tablets) = 0.4 x 250 = $100.00. Process costs, that include material direct labor, energy, and overhead costs = $400. With a process yield rate of 94%, the total cost per 1000 acceptable tablets = $400/0.94 = $425.53, which yields the cost/tablet of acceptable product = $0.43.

b) With an improved yield of 96%, the cost/tablet of conforming product = $400/0.96 = $416.67/1000 tablets = $0.42/tablet. The relative level in capacity = 0.96/0.94 = 1.0213, indicating a 2.13% increase in capacity.

c) New labor costs = $85/1000 tablets, and new energy costs = $40/1000 tablets. Total process costs now = $150 + $85 + $40 + $94 = $369/1000 tablets. Assuming the process yield to be 96%, the cost per 1000 acceptable tables = $369/0.96 = $384.38, yielding a cost/tablet of conforming product = $0.38. The percentage reduction in cost from the original process = (425.53 – 384.38)/425.53 = 9.67%.

1-19. a) Total cost of goods sold, including marketing costs, = $(20 + 30 + 6 + 25 + 25 + 10) = $116/m^3. Assuming a 100% first-pass yield, for a 10% profit margin over cost of goods sold, the selling price = $127.6/m^3.

b) With a first-pass yield of 94%, cost of goods sold for conforming product = $123.40/m^3. With the selling price being the same as in part a), the profit = $(127.6 - 123.4) = $4.20/m^3. So, profit margin as a proportion of cost of goods sold = 4.20/123.40 = 3.40%.

c) With a first-pass yield of 98%, cost of goods sold for conforming product = $116/0.98 = $118.37/m^3. If the sales price is kept at $127.6/m^3, the unit profit = $9.23/m^3. The profit margin as a percentage of cost of goods sold = 9.23/118.37 = 7.80%.

d) The additional capital expenditure = $150,000 and the demand rate is 5000 m^3 monthly. With a profit of $9.23/m^3, the volume of sales required to break even = 150000/9.23 = 16,251.35 m^3. Hence, the time to break even = 16251.35/5000 = 3.25 months.

e) The improved process has a first-pass yield of 98%, meaning that for 100 m^3 of production, 98 m^3 is conforming and 2 m^3 is of lower quality. For conforming product, profit = $(127.60 - 116) = $11.6/m^3. For lower quality product, profit = $(120-116) = $4/m^3. Using the concept of weighted average, profit = 11.6 (0.98) + 4.0 (0.02) = $11.448/m^3. The break even volume now = 150000/11.448 = 13102.725 m^3. Thus, the time to break even = 13102.725/5000 = 2.62 months.

1-20. The flow chart of the four operations is shown in Figure 1-1, with the yield, unit processing cost, and unit inspection costs indicated:

a) The first-pass yield at the end of four operations = (0.95 (0.90) (0.95) (0.85) = 0.6904. Total processing costs per unit = 10 + 6 + 15 + 20 = $51. Hence, the unit cost for conforming product = 51/0.6904 = $73.87.

b) Inspection is now conducted after the first and second operation: It is assumed that the inspection process correctly identifies all parts and nonconforming parts are not forwarded to the next operation. Using the first-pass yields, processing costs per 1000 parts = 1000 (10) + 950 (6) + 855 (15) + 855 (20) = $45,625. Inspection costs per 1000 parts = 1000 (0.50) + 950 (2) = $2400. Hence, total costs for processing inspection for 1000 parts = $48,025. The number of conforming parts for 1000 parts produced = 1000 (0.6904) = 690.4. Hence, the unit cost per conforming part = 48025/690.4 = $69.56.

c) We now consider the case where inspection is conducted only after the third operation: Using the first-pass yields, processing costs per 1000 parts = 1000 (10) + 1000 (6) + 1000 (15) + 1000 (.95) (.90) (.95) (20) = $47,245. Inspection costs per 1000 parts = 1000 (3) = $3000, yielding total processing and inspections costs per 1000 parts = $50,245. Hence, the unit cost per conforming part = 50245/690.4 = $72.78.

d)   In general, it is desirable for inspections to be conducted in the early operations in the process, if one has to choose between locations of inspection. If inspection is conducted early on in the process, it will hopefully eliminate nonconforming product from going through subsequent processing and incurring these costs. Also, for operations that have expensive unit processing costs, it is desirable to conduct inspection before such processing so that nonconforming product can be eliminated prior to processing.

1-21.   Total cost function is given by TC $= 50q^2 + 10q + (5 + 85)(1-q) = 50q^2 - 80q + 90$. This function can be plotted as a function of the quality level, q, to determine the value of q where the total cost is minimized. Alternatively, taking the derivative of the total cost function with respect to q, we obtain, $(100q - 80)$. Equating this derivative to 0, yields the operational level of quality for this static situation as $q = 80/100 = 0.80$. It can be seen that the total cost per unit at this level of quality is TC $= 50(0.8)^2 - 80(0.8) + 90 = \$58.00$.

The form of the cost functions assumed are as follows. Prevention costs increase quadratically as a function of q, while appraisal costs increase linearly with q. It is possible that appraisal costs might actually decrease with the level of q. Further, it is possible that, beyond a certain level of quality, the combined prevention and appraisal costs might show a decrease in the rate of increase as a function of q. Internal failure and external failure costs are assumed to be linearly decreasing as a function of q. Here again, such cost functions might decrease as a non-linear function of q, with the rate of decrease diminishing with an increased level of q. When all such considerations are taken account, it is possible that the desirable operational level of quality is towards a goal of 100% conformance. In this situation, the total cost function may decrease as a function of q, rather than show the traditional u-shaped form.

1-22.   The total cost function, per unit, as a function of the quality level, q, is given by:

$$TC = 50q^2 + 2(1 - q) + 5\frac{(1 - q)}{q} + 85\frac{(1 - q)}{q}$$
$$= 50q^2 - 2q + \frac{90}{q} - 88.$$

Figure 1-2 shows a graph of the total cost function as a function of q.

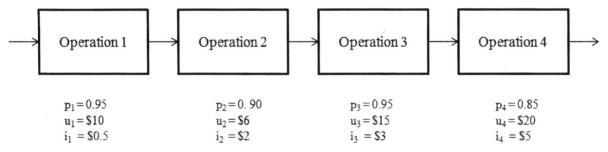

Operation 1	Operation 2	Operation 3	Operation 4

$p_1 = 0.95$ $p_2 = 0.90$ $p_3 = 0.95$ $p_4 = 0.85$
$u_1 = \$10$ $u_2 = \$6$ $u_3 = \$15$ $u_4 = \$20$
$i_1 = \$0.5$ $i_2 = \$2$ $i_3 = \$3$ $i_4 = \$5$

Figure 1-1. Operations Sequence and Unit Costs

13

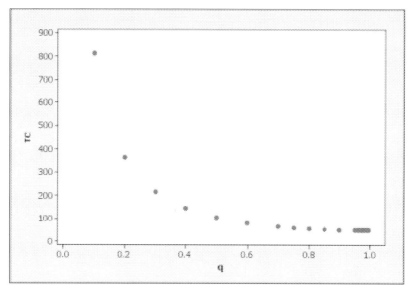

Figure 1-2. Plot of total cost versus quality level

From the total cost function, the level of q that minimizes the total cost can be identified. This is found to be approximately $q \simeq 0.975$.

Alternatively, the derivative of the total cost function may be found and equated to zero. We have:

$$100\, q - 2 - \frac{90}{q^2} = 0$$

or $100\, q^3 - 2q^2 - 90 = 0$, which yields $q \simeq 0.975$.

1-23.   The revenue function is given by $90q^2$. So, the net profit function is expressed as:

$$\text{NP} = 90q^2 - (50q^2 - 2q + \frac{90}{q} - 88) = 40q^2 + 2q - \frac{90}{q} + 88.$$

This function may be plotted as a function of q, and the level of q that maximizes this function can be determined. Alternatively, the derivative of the net profit may be found and equated to zero. We have:

$$80q + 2 + \frac{90}{q^2} = 0 \text{ or } 80q^3 + 2q^2 + 90 = 0, \text{ which yields } q \simeq 1.000, \text{ implying that total}$$

conformance is the desired option.

14

CHAPTER 2

SOME PHILOSOPHIES AND THEIR IMPACT ON QUALITY

2-1. There are several customers in health care and some of their needs are described as follows:

Patients - Desire excellent clinical treatment, good customer service, affordable cost.

Physicians - Desire excellent clinical quality, good facilities, good support staff and nurses, acceptable schedule.

Nurses - Desire good work environment, acceptable schedule, support for professional development.

Government - Desire adequate health care coverage of all people, affordable cost of coverage, mechanism to monitor quality of health care and medication.

Insurance Companies - Desire payment on-time, good relationships with healthcare organizations to determine acceptable reimbursement policies, improve market share.

Employers - Desire availability of skilled labor, availability of adequate physicians and nurses, funds to update facilities, support from local government and community.

Employees - Desire good work environment, job satisfaction, opportunities for professional development.

Community - Desire affordable health care, easy access to facilities.

Families and friends of patients - Desire prompt and effective treatment, service-oriented mentality of staff and personnel, convenience of visits to the facility.

2-2. a) Health care - Incorrect results from laboratory test, incorrect diagnosis, discourteous treatment by admissions staff or nurse.

b) Call center - Incorrect routing of call, incorrect information provided by operator, long holding time before call being answered, discourteous conversation by operator.

c) Internal Revenue Service - Error in processing tax return, long delay in responding to question from individual, wrong information provided by customer representative, discourteous behavior in answering customer inquiry.

d) Airline industry - Delay in arrival time at destination, lost baggage, discourteous behavior by customer check-in agent or stewardesses.

2-3. Health Care - Incorrect results from laboratory or incorrect diagnosis should be easily available from documents. Discourteous treatment is not always measurable. If patients are asked to complete a survey on conclusion of their stay, with adequately designed surveys, it could be measured, usually on an ordinal scale (say 1-10).

Call center - Long holding time could be measured based on setting up equipment that documents when incoming call was received and the time it was answered. Incorrect routing of call or incorrect information is more difficult to identify and measure, unless there are provisions of feedback from the customer. Perhaps, a web-based survey could ask questions pertaining to these issues. Discourteous treatment may be recorded from survey feedback or customer complaints – not as easily obtainable.

Internal Revenue Service - The government agency should be able to easily identify errors in processing, specially if brought to their notice by the customer. Long delays in response may be observed if proper records/electronic documentation are maintained. Discourteous behavior may be identified through review of recorded conversations, that are normally used to train/monitor performance of customer representatives.

Airline industry - Delay in arrival time and lost baggage information can be easily obtained. Discourteous behavior is more difficult to measure since customers report such cases only in extreme situations. Customer surveys could possibly be used to collect such information.

2-4.   Health care - Perception of quality influenced by factors such as national or regional rating of health care facility/physician, certification by accreditation agencies such as JCAHO, education and prior experience of physicians/nurses, HMOs that are associated with the facilities/patients, and cost of service relative to their comparable facilities, among others. Several of these could be managed through appropriate disclosure of the quality level of operation, pursuing and maintaining accreditation through professional organizations, participation of physicians in national/international meetings of cohorts, appropriate advertisements, keeping up with the changing needs of the patient through the promotion of well-ness activities and awareness of potential health risks.

Call center - Customer perception of quality influenced by factors such as image depicted by media, advertisements displayed in print or electronic media, product/service options offered, relative cost of product/services with respect to competitors, degree of warranty coverage relative to that of competitors, and upgrading product/service options based on evolving needs of the customer, among others. Much of the media advertisements can be managed by the company. The company can definitely make an effort to survey the changing needs of the consumer and create product/service options accordingly to meet such needs.

Internal Revenue Service - Too often, the answers to questions posed by a customer are rather vague and general. They lack specificity, which influences customer perception of quality. The series of steps a customer goes through, before getting a customer representative on the telephone, diminishes perception of quality. All of these can be addressed by the IRS through adequate training as well as implementation of additional customer service lines. The web can be utilized to upgrade the quality of response to certain questions.

Airline industry - Comparison data on airline performances, made available to the public, influences customer perception of quality. The proportion of delays, the average delay, the proportion of bags mishandled are example of statistics that are published by agencies. Customer satisfaction ratings on services are also reported by consumer agencies. Companies can definitely manage the majority of these factors. Accuracy and speed of checking in passengers, adequate bar-coding on bags to prevent mishandling, attention to detail during in-flight services are measures that could be easily incorporated by the company.

2-5. a) Stratified probability sampling where samples could be selected from people with income in an upper level or assets in an upper level. These products are usually purchased by people in a high income bracket.

    b) Simple random probability sampling from the population of households since such a product is a common household item.

    c) Simple random probability sampling from the population since cell phones are common among all people.

    d) Judgment sampling since boutique clothes are usually purchased by people with a certain taste. The expert should identify the common characteristics of people who make such purchases, which may then be used as a basis to draw samples. Alternatively, if characteristics of such buyers are easily identifiable, perhaps the population can be stratified based on these characteristics. Then, a cluster sample could be selected from the desired stratum.

    e) Since all households subscribe to city municipal services, a simple random sample from the population of households could be chosen.

    f) Stratified probability sampling where incomes of individuals are stratified into strata or groups. Next, simple random samples could be chosen from within a stratum. This will ensure representation of individuals from quite distinct income groups in the sample.

    g) Stratified probability sampling where home values in the population are stratified. Since coverage of very expensive homes are quite different from that of average homes, creating strata will help in identifying such differences. Next, simple random samples could be chosen from within a stratum.

2-6. a) The extended process includes not only the organization but also the customer, vendors, investors or stakeholders, and the community in which the organization resides. By including these entities in the extended process, emphasis will be placed on satisfying the customer as well as optimizing the total system. For an organization that assembles computers, vendors will consist of those supplying parts and components, for example, those that provide the motherboard, the various drives and readers/writers, the monitor, the casing, and so forth. Investors may consist of shareholders. The community will include the locality which provides the source of employees and where the facility resides. As the company does well and market share grows, it may have an impact on the local economy.

b) For a hospital, the primary customers are the patients. Secondary customers are physicians, nurses, technical staff, employees, federal or state government, health-maintenance organizations, and insurance companies. Vendors include those who sub-contract and provide food services, linen services, parking services (if not part of the hospital), pharmaceutical companies, companies that provide X-ray equipment, and so forth. Hospital administration and shareholders could be a part of the investor group. Friends and families of patients and the local community that is served by the hospital are part of the extended process.

c) For a software company, primary customers are people/organizations that place an order for the software. Secondary customers consist of employees, technicians, and managers. If parts/segments of the software are provided by third parties, they are the vendors. Intellectual property rights could be a matter of discussion. Sometimes, for open software, just an acknowledgement might be sufficient. Time to develop software and its accuracy could be influenced by technical personnel within the company as well as third-party vendors. The extended process that includes highly motivated technical persons, who write code, could actually influence the development of a better product in a shorter period of time.

d) In the entertainment industry, say the movie industry, the primary customer would be the movie-goer. However, other customers include people who rent DVDs of the movie or use the available-on-demand option through a cable company. Secondary customers include the movie production group and actors, editors, choreographers, etc. Those associated with promotion and advertising for the movie could fall in this group, if they are part of the parent company. When some of the movie production functions or advertising functions are sub-contracted, they become the vendors. The financiers are part of the stakeholders.

2-7. Inspection is merely a sorting mechanism that separates the nonconforming items from the conforming ones. It does not determine the causes of nonconformance to identify appropriate remedial actions. Hence, it is not a viable alternative to quality improvement. Depending on mass inspection to ensure quality does not ensure defect-free product. If more than one inspector is involved in mass inspection, an inspector assumes that others will find what he/she has missed. Inspector fatigue also causes defective products to be passed. The fundamental point is mass inspection does not improve or change the process and so it is not an alternative to quality improvement.

2-8. a) Some general principles of vendor selection involve consideration of the attributes of price, quality, ability to meet due dates, flexibility to adapt to changes in the company's product, and ability to meet demand fluctuations, among others. The idea is to build a long-term partnership with the vendor who becomes a part of the extended process.

For a supermarket store, meeting delivery dates is quite important since replenishments are often made multiple times a week. The supermarket depends on its vendor for its supplies of produce and meat, both perishable products. It is important, therefore, for the supplier to deliver a quality product. Demand for certain items are seasonal, for example watermelons during summer. The supplier must have the ability to deliver such seasonal items, often in increased quantities during peak demand. Non-perishable goods are replenished on a stock-to demand basis. Price of goods offered by the vendor must be on a competitive basis.

b) For a physician's office, supplies that are replenished by vendors may include routine items necessary for patient examination, for example latex gloves, replacement paper cover for patient examination units, syringes, and other hygiene products. The suppliers are selected based on their ability to meet due dates and volume. Medical equipment such as stethoscopes or blood pressure measuring equipment require purchasing reliable equipment that are competitive on a price and warranty basis. Moreover, other heavy equipment such as dialysis machines, X-ray units, etc. require a competitive price and high quality as offered by the vendor and demanded by the company.

c) For a fast-food restaurant, the basic ingredients that include meat, buns, salad ingredients, and fruit have to be replaced routinely. They must be fresh and of choice quality, with the supplier being dependable. Timely delivery is of utmost importance. For other items replenished less frequently, such as cooking oil or frozen food (such as onion rings or potato fries), price and quality are important.

2-9. Organizational barriers prevent or slow down the flow of information internally between departments or between the employee and the supervisor. Among external barriers, these include the flow of information between the company and its vendors, the company and its customers, the company and its investors, and the company and the community in which it resides.

To improve communication channels, there needs to be a change in the organizational culture of the company. Free and open channels of communication needs to be established by management. There should be no punitive measures or repercussions to employees who provide feedback or products/processes, with employees being able to express their opinions honestly. Management can demonstrate this only by example or through implementation of such practices.

A second approach could be to promote a team effort in improving products/processes. While individual skills are important, a variety of persons are involved with multiple operations in making the product or rendering the service. It is the joint impact of all these people that influences quality. The reward structure, created by management, could be established in terms of the output quality of the team unit.

The adoption of cross-functional teams to identify product/process changes for quality improvement will definitely promote open channels of communication and reduce existing barriers. A system to accept suggestions from employees at all levels (including managerial personnel) could also be adopted by senior management. Further, a system that rewards the person/team when a proposed idea is implemented, will definitely boost morale and provide an inducement for fresh ideas.

2-10. The traditional performance appraisal system does not distinguish between inherent variation in the evaluation of employees. Variations exist between those conducting the evaluation. Thus, if the same employee were to be evaluated by these evaluators, their assigned ratings could be different. Secondly, for employees who are part of the same "system", there are no significant differences, even though they may be assigned different ordinal numbers indicating a degree of relative performance. These conclusions are based on the assumption that performance of individuals in a system varies according to a normal distribution. Therefore, rating categories that fall within the bounds of this system (say within three standard deviations of the mean), are not significantly different, statistically. Those outside these bounds are considered significantly different. For personnel that fall on the upper tail indicating "outstanding" performance relative to the others, merit pay/incentives/rewards must be provided. Likewise, for those personnel that fall on the lower tail indicating "poor" performance, suggestions/plans through which they may improve their performance should be provided.

Management has the responsibility of designing an adequate performance appraisal system. The system should reward outstanding performance and should provide specific guidelines to those whose performance is deemed unsatisfactory to improve their performance. Management, in this context, must also provide specific means for professional development. Another criterion for management consideration is to design a performance appraisal system that promotes teamwork. Since it is the combined effort of a group of individuals that results in an excellent product or outstanding service, members of the team with outstanding performance should be rewarded appropriately.

2-11. Quality control deals with identification of the special causes, determining remedial actions, and implementing these actions so that the special causes are eliminated from the system. These are sporadic in nature. Frequently, the remedial actions could be determined at the operator level or lower line management level. Quality improvement, on the other hand, deals with identification of common causes that are inherent to the system and determining appropriate actions to eliminate or usually reduce their impact on the variation in the product/service. These decisions are usually made at the management level and involve changes in the system. They require decisions on resource allocation that usually are not made at the operator/lower management level. For example, replacement of major equipment for processing in order to reduce variation is an item of quality improvement. Alternatively, use of an incorrect form that caused a delay in order processing, could be an issue of quality control. Usually, quality control issues are handled first, followed by quality improvement.

2-12. Deming's first deadly disease is management by visible figures only. Visible figures, such as monthly production, do not often demonstrate the true state of affairs, as the amount of rework and scrap may not be captured. Certain measures such as employee satisfaction, loss of market share due to customer dissatisfaction, and loss of goodwill are difficult to measure by numbers. Management must create a climate of innovation and trust, drive out fear, promote teamwork, and develop a participative atmosphere.

Lack of constancy of purpose is Deming's second deadly disease. A mission statement that is viable and conveyed to all employees is imperative. This should determine the long-term vision of the company. There must be commitment to this mission, through visible means, by management. Examples could be commitment of resources, establishment of a quality culture that promotes the concept of continuous quality improvement, and focus on process improvement.

The third deadly disease is performance appraisal by numbers. Some drawbacks of such an appraisal has already been previously discussed (question number 10). An appraisal system must be created by management that differentiates between statistically significant differences in performance. Inherent variation that exists between appraisers must be accounted for. Further, the concept of synergism through effective teamwork must be integrated into the system. An appraisal system that rewards teamwork should be developed by management.

Deming's fourth deadly disease is a focus on short-term orientation. When the focus is on short-term profits or short-term return on investment, it causes actions that are counterproductive to long-term stability and growth of the company. Much-needed investment on personnel development and equipment replacement may be postponed, leading to lack of a competitive edge in the long term. Along these lines, vendor-vendee relationships should be cultivated so that they are sustained on a long-term basis and both are part of the extended process.

Finally, mobility of management is Deming's fifth deadly disease. With frequent changes in management, the constancy of purpose of the company is lost. Senior management must demonstrate, through actions, their commitment to other lower levels of management through their support of professional development, opportunities to provide input to strategic decisions, support for continuous quality improvement of products and processes, and concern for the welfare of the manager through an adequate reward and recognition system.

2-13. Several of the drawbacks of a bottom-line management approach are elucidated in the response to the five deadly diseases in the context of Deming's philosophy. Placing an emphasis only on, say measures such as total revenue or net profit, on a short term basis, could be counterproductive to actions that are desirable for long-term stability and growth. It may deter innovative approaches to product/process design since such actions may require substantial capital investment initially. While the positive impact of such investment may not be felt in the short-term (for example, quarterly profits), in the long-term it may lead to an extremely competitive position within the industry. Such an approach may also deter quality improvement activities, since return on investment may not be realized in the short-term.

2-14. The answers to question numbers 12 and 13 address the issues between short-term profits and long-run stability and growth. The impact of capital investments are usually not realized in the short-term term. So, when such decisions are made, figures on short-term profits may not be attractive. However, such decisions may create a competitive edge of the company in the long run. Management must focus on long-term stability and growth and constantly seek innovations in their products and processes. This is the only way that they can maintain or improve their relative standing.

2-15. a) Deming's system of profound knowledge is explained in the context of a hospital. The first principle is based on knowledge of the system and the theory of optimization. The extended system consists of patients, physicians, nurses, government, insurance companies, employees, vendors that provide services to the hospital, shareholders, and the community that includes families and friends of patients. So, optimizing only the availability of nurses (a sub-system within the total system), may not be the desirable solution in the context of the total system. The second principle relates to knowledge of the theory of variation. Variation is due to special causes and common causes. Special causes are external to the system while common causes are inherent to the system. Management has the responsibility to address common causes. Thus, a delayed X-ray report could be due to a special cause created by an experienced technician being absent on a given day. On the other hand, an extended length of stay for patients in a certain diagnosis related group, could be due to the current processes and procedures that exist in the hospital. A thorough investigation of such practices may be necessary to address the common causes in the system.

The third point refers to exposure to the theory of knowledge. Predictions on the system are made based on a set of assumed hypotheses. Through observed data, the hypotheses are validated. In this context, after study of the hospital practices, we may hypothesize that an extended length of stay occurs due to a practice where laboratory results are received after 4 pm. Due to the lateness of receiving such results, even if they are satisfactory, there is not sufficient time to process the patient's discharge that particular day, thereby extending the length of stay. Perhaps, management could institute a change in the policy whereby test results should be delivered by 2 pm, making it possible to discharge a patient that same day. The fourth point deals with knowledge of psychology. Motivating people requires a knowledge of intrinsic and extrinsic factors. In this environment, nursing staff could be motivated by salary and reasonable working hours. Physicians could be motivated by acceptable schedules, availability of skilled nurses, and availability of upgraded cutting-edge equipment required for surgical procedures.

b)　In the context of a software company, the primary customers are individuals, businesses, corporations, or government organizations who place an order or purchase the software. Secondary customers, within the extended process, may include software design and coding personnel, vendors who produce the software in a given media (i.e. CD), marketing and advertising personnel, software support personnel, external code developers, investors, and community in which the company is located. Meeting and exceeding the needs of these various entities will help in the achievement of creating a win-win situation for all constituencies. Optimization of the system must consider meeting these various needs.

Both special and common causes may exist in the system. Unusual delays in responding to customer problems by support personnel could be due to special causes, such as incorrect routing of call to wrong person. On the other hand, if such delays are due to lack of training of personnel, which is a system-related common cause, management should decide on providing appropriate training.

The ability to make predictions using the theory of knowledge could be used to determine the amount of memory space required to run programs to applications of a certain defined size. This will enable the user to determine the details and configuration of hardware requirements. Knowledge of psychology to motivate people will be influenced by the type of personnel that we focus on. For instance, to motivate people who write code, money is not the only motivation. To solve a challenging problem in a computationally-efficient manner may provide the impetus. Provision of adequate work space with access to recreational facilities could be motivating. However, for software-support personnel, an acceptable work schedule and adequate salary could be rewarding.

2-16.　Some organizational culture issues that management must address in order to strive for long-run stability and growth are as follows:

i)　Create a culture that moves away from short-term results such as quarterly profits or quarterly production. In order to accomplish this, a change in the reward structure must take place, where employees will be rewarded based on their innovativeness and contributions to quality improvement.

ii)　Create a culture of long-term partnerships with vendors. Integrate vendors into the system such that product changes are automatically conveyed to vendors concurrently. Provide technical support to vendors, if necessary, such that they are prepared to implement/incorporate product design changes.

iii)　Improve transmission of information to the various constituencies on a timely basis. This may require a change in the culture that does not treat suppliers and customers as part of the extended system. Note that customer needs may change on a dynamic basis. Hence, information flow of such needs must be collected on a timely basis too.

iv) Adopt a culture that promotes teamwork and overall improvement in the product/process rather than just individual operations.

v) A culture of continuous improvement must be adopted to strive for excellence on an on-going basis. This entails striving to become the benchmark for the industry.

vi) A climate of mutual respect, where fear has been driven out, must be created. People perform at their peak when they are given ownership of the product/process.

vii) A mission must be adopted that is understood by everyone in the organization. Management must demonstrate their commitment to the mission. Input must be sought from all levels.

2-17. a) The airlines reservation system based on the concept of yield management attempts to manage the supply and demand side simultaneously in a dynamic manner. The concept is based on partitioning demand based on categorizing customers: Business travelers who will buy seats at the full price, tourists who are looking for competitive price deals (relative to those offered by other airlines), and those using accumulated sky miles. The total supply of seats on a given flight is fixed. However, the airlines can manage its allocation of supply of seats in each of the above categories, based on data on demand that is collected on a real-time basis.

Customer satisfaction will be influenced by the "category" of the customer since priorities may be different for each category. For the business traveler, the important factor is getting a seat at the last minute where price is not a consideration, with the company reimbursing the cost. For the tourist, being able to obtain a competitive ticket price for the chosen date/time are important. For the person using sky miles, satisfaction occurs when a seat is available for a desired date/time.

b) It could lead to customer dissatisfaction if price is the only factor based on which satisfaction is derived. Note that even for those in the tourist class, since seat prices change on a dynamic basis, the price paid for a seat will be different based on the day when the ticket was purchased.

c) The objective function is dynamic in nature and is influenced by the demand distribution in each category. These demand distributions are projected based on historical data for the same route and same time frame, accounting for "special" causes that may impact demand. The model also requires estimation of certain costs, such as the cost of overselling, in which case the customer has to be booked on the next flight and could be awarded certain amount of dollars towards purchase of a ticket within the next year. It requires estimation of the cost of underselling, with a seat going empty. In this case, however, there are various factors of influence. For example, if seats allocated to the "business" category were shifted to the "tourist" category, the company loses potential revenue from a business traveler who would have paid full fare. Likewise, there is a different opportunity cost if there is an unfilled demand for a "tourist" category seat, which had been shifted to a "business" category and goes empty.

d) Such yield management practices could be used in other industries that have capacity restrictions, demand is dynamic and could be influenced by price, and excess capacity on a given date is lost. For instance, the hotel and hospitality industry is an example. An on-line reservation system could be set-up that monitors prices based on demand distribution and rates offered by competitors. The hotel industry has an alternative way of increasing capacity in the event of overbooking. They will transport their guests to another facility (with whom they subcontract) and put them up free of charge. Rental car companies could be another example that could utilize such yield management practices. Here again, the capacity is limited, at a given site. Demand fluctuates, but customers could also be categorized. For example, to the business traveler, price is not a concern, rather availability and the type of car desirable. For the tourist, price could be the main criterion, followed by desired size or type (full size, fuel efficient). A third example could be cruise lines, that have similar features to the airline industry, with competition being not as extensive. Demand is usually seasonal, providing opportunities for the company to offer attractive off-season rates. Movie theatres could also utilize such concepts to simultaneously manage fixed supply with demand. For instance, reduced prices could be offered to matinee or shows prior to 6 pm.

2-18. American Express could utilize the database of information it has on its card holders' spending habits. Knowing the types of goods/services that a given customer purchases, it could provide information on special offers, seasonal offers, or promotional offers on related goods/services. It could negotiate agreements with manufacturers of products or providers of services to obtain either better warranty coverage or even better rates for its customers, thereby creating an avenue to improve customer satisfaction. An example of this could be in providing complete travel services to its customers, which could include airlines, hotel, tour options/price, and car rental, all as part of a complete package. The customer could be given options to select from the package.

From the annual summary of database of spending habits, by providing information to retail outlets, such stores could determine amounts to stock of the various types of goods. The retail stores could also determine suitable pricing policies based on spending patterns. They could determine related items to stock, that would be of interest to the particular customer. For a travel company, it may assist in the determination of types of packages to offer.

CHAPTER 3

QUALITY MANAGEMENT: PRACTICES, TOOLS, AND STANDARDS

3-1.   There are three major themes in the total quality management philosophy – customer, process, and people. Satisfying and exceeding the needs of the customer is the foremost objective. The core values of the company are management commitment and a directed focus of all employees towards a common vision and mission. Senior management creates a strategic plan, while mid-management develops operations plans accordingly. Implementation of plans requires an organizational culture that empowers people to suggest innovations through open channels of communication. Further, focus is on process improvement, where suppliers, customers, and investors are integrated into the extended process. The selection of a company will vary with the individual. Each should identify the particular quality culture of the selected company and the manner in which it fits the general themes previously discussed.

3-2.   A long-term partnership with the vendor creates a sense of extended relationship, with the company able to assist the vendor in a variety of ways. For issues dealing with product changes, the vendor is automatically notified (sometimes ahead of time, when known) so that appropriate process changes in the vendor operations can be implemented with the least delay. Further, fluctuations in demand that may lead to increases during peak seasons could be communicated to the vendor ahead of time, so that there is no increase in lead time of obtaining parts or components. This ensures delivery to the customer to be on time as well as meet desired quantity. Additionally, when the company and the vendor work together in addressing process-related problems or quality improvement activities, there is a synergy that may lead to cost and/or time reduction.

3-3.   Depending on the selected hospital, the vision, mission, and quality policy will vary. A vision is based on what the hospital wants to be. It could be, "Become the hospital of choice for people in the region." A mission statement, derived from the vision, is more specific and goal-oriented. It could be, "Achieve a high level of patient satisfaction, from in-patients and out-patients, through the best clinical care and motivated employees." A quality policy is a road map that indicates what is to be done to achieve the stated vision and mission. This could be, "Obtain necessary feedback from patients and the hospital staff (that includes physicians, nurses, and technical staff) to identify causes of action that support continuous quality improvement."

3-4.   Motorola's concept of six-sigma quality, even though it may have started out as a metric for evaluation of quality (say parts per million of nonconforming product), could be viewed as a philosophy or as a methodology for continuous improvement. In terms of a metric, Motorola's assumption is that the distribution of the quality characteristic is normal, and that the process spread is much smaller than the specification spread. In fact, it is assumed that initially, the specification limits are six standard deviations from the mean. Subsequently, shifts in the process mean may take place to the degree of 1.5 standard deviations on a given side of the mean. Here, the assumption is that larger shifts in the process mean will be detected by process controls that are in place and corresponding remedial actions will be taken. Thus, the nearest specification limit is 4.5 standard deviations from the mean, while the farthest specification limit being 7.5 standard deviations from the mean, after the process shift. Using normal distribution tables, it can be shown that the proportion of nonconforming product (outside the nearest specification limit) is 3.4 parts per million. The proportion nonconforming outside the farthest specification limit is negligible, yielding a total nonconformance rate of 3.4 ppm.

As a philosophy, the six sigma concept is embraced by senior management as an ideology to promote the concept of continuous quality improvement. It is a strategic business initiative, in this context. When six sigma is considered as a methodology, it comprises the phases of define, measure, analyze, improve, and control, with various tools that could be utilized in each phase. In the define phase, attributes critical to quality, delivery, or cost are identified. Metrics that capture process performance are of interest in the measure phase. In the analyze phase, the impact of the selected factors on the output variable is investigated through data analytic procedures. The improve phase consists of determining level of the input factors to achieve a desired level of the output variable. Finally, methods to sustain the gains identified in the improve phase are used in the control phase. Primarily, statistical process control methods are utilized.

3-5.   Quality function deployment has the advantage of incorporating customer needs or requirements into the design of a product/service. The approach prioritizes customer needs and identifies specific means in the product/process design that helps in meeting such needs. Out of several alternative proposals, based on the priority ranking of needs and the relative impact of each alternative on meeting each customer need, a weighted index is developed. Under a scarcity of resource environment, proposed alternatives are selected based on the computed weighted index. QFD reduces product development cycle time through consideration of design aspects along with manufacturing feasibility. It also cuts down on product developmental costs through consideration in the design phase of the myriad of issues that deal with technical ability of the company relative to competitors. There are some key ingredients necessary for the success of QFD. First, a significant commitment of time has to be devoted to complete the QFD process. Second the use of cross-functional teams is a necessary mode for information gathering required for the QFD process. While this is a significant resource commitment, the advantage is it leads to an improved product/process design.

3-6.   The selection of an organization will influence the type of strategy that is adopted and the associated diagnostic and strategic measures in each of the four areas of learning and growth, internal processes, customer, and financial.

   a)   Information technology services – Learning and growth perspective: Some diagnostic measures are exposure to recent developments in software and technical knowledge in the field. Some strategic outcome measures are retention or attraction of skilled technical people to the company and employee satisfaction. Some strategic performance measures could be policies to empower employees that support innovation and prevalent reward structure. Internal processes perspective: Some diagnostic measures are errors per 1000 K lines of code, type of coding errors and their severity, and delay in responding to customer requests. Some strategic outcome measures are level of service to the client (who could be internal or external), and time to develop software and implement it. Some strategic performance measures could be absenteeism rate of employees and efficiency of employees. Customer perspective: Some diagnostic measures are time to solve and respond to customer problems and number of monthly customer complaints. Some strategic outcome measures are cost of providing service, reliability of operations, and degree of customer satisfaction. Some strategic performance measures could be cost of subcontracting certain services, and degree of trust in relationship with vendor. Financial perspective: Some diagnostic measures are changes in unit cost of providing services, and costs due to not meeting warranty/liability obligations. Some strategic outcome measures are return on investment, and market share of company. Some strategic performance measures could be degree of investment in equipment and infrastructure, and operating expenses.

b)   Health care – Learning and growth perspective: Some diagnostic measures could be lack of timely feedback by nurses, and delays in admitting patients due to incorrect information provided by admitting staff. Some strategic outcome measures are degree of physician and nurse satisfaction, and number of suggestions for improvement by laboratory personnel and admissions staff. Some strategic performance measures could be the type of reward structure for physicians or nurses, and incentive schemes for hospital staff. Internal processes perspective: Some diagnostic measures are average time to process in-patients for admission, delay in processing an X-ray, and medication errors per patient-day. Some strategic outcome measures are readmission rate, and length of stay for a certain diagnosis related group. Some strategic performance measures are infection rate and blood culture contamination rate. Customer perspective: Some diagnostic measures are time to discharge patients, and time to deliver patient from check-in at the emergency department to a hospital bed. Some strategic outcome measures are degree of satisfaction by in-patients, and proportion of patients that would recommend others. Some strategic performance measures are cost of a certain surgical procedure, and treatment of patient by nursing staff. Financial perspective: Some diagnostic measures are proportion of reimbursement requests denied by Medicare, and physician time in surgery lost due to inappropriate scheduling. Some strategic outcome measures are cost per case of in-patients with a certain diagnosis, and market share captured. Some strategic performance measures could be percentage of asset utilization of operating room capacity, and reduction in unit costs of laboratory work.

c)   Semiconductor manufacturing – Learning and growth perspective: Some diagnostic measures could be long inspection time of sampled product, and long set-up time of equipment due to lack of skill. Some strategic outcome measures could be number of process improvement suggestions received from employees, and reward structure to promote environment of continuous improvement. Some strategic performance measures could be spending on employee professional development, and type of recognition system, beyond pay, available to staff. Internal processes: Some diagnostic measures are set-up time for equipment due to changes in product specifications, and lead time for delivery of raw material/components due to change in customer requirements. Some strategic outcome measures are time to develop new process based on a new product innovation, and total cost per batch (1 million) of microchips. Some strategic performance measures could be expenditures in research and development of processes, and unit procurement costs from vendor. Customer perspective: Some diagnostic measures are response time to meet changes in customer orders, and number of shipments rejected by the customer. Some strategic outcome measures are proportion of customers complimentary of the company, and increase in annual referrals by customers. Some strategic performance measures could be time to serve customers with a certain minimum volume of orders, and degree of discount offered to customers with high volume of orders. Financial perspective: Some diagnostic measures are overhead costs per batch and cost of machine downtime per month. Some strategic outcome measures are return on investment and growth in market share. Some strategic performance measures could be percentage of equipment utilization and amount of investment to upgrade equipment.

d) Pharmaceutical company – Learning and growth perspective: Some diagnostic measures could be proportion of proposals rejected due to lack of technical competency of staff, and time to develop a proposal for consideration of senior management. Some strategic outcome measures are degree of satisfaction of technical staff, and revenue per employee-hour. Some strategic performance measures are incentive plans for scientists, and number of successful proposals annually. Internal processes perspective: Some diagnostic measures are time to develop a batch of prototype, and throughout rate of a prototype. Some strategic outcome measures are cost per batch of tablets, and proportion nonconforming (ppm) of product. Some strategic performance measures are proportion nonconforming (ppm) of shipments from vendor and unit overhead costs per batch. Customer perspective: Some diagnostic measures are time to conduct survey of proposed drug, and lead time to meet customer order changes. Some strategic outcome measures are percentage of satisfied scientists and engineers, and proportion of senior personnel retained. A strategic performance measure could be time to meet a competitor's deadline for a new product development. Financial perspective: Some diagnostic measures could be cost of product that is scrapped due to not meeting desired specifications, and overhead costs per batch. Some strategic outcome measures could be profit margin, and sales growth. Some strategic performance measures are cost savings due to equipment changes, and investment in equipment.

3-7. For the airlines industry, customer requirements could be as follows: Price of ticket, convenience of schedule, delay in arrival, lost baggage, and in-flight service, among others. Based on customer survey, priority ratings may be applied to the above requirements. Some technical descriptors could be as follows: Select cities to serve based on competition and demand, type and size of fleet, baggage identification (barcoding) and handling procedures, training of in-flight attendants and provision of desirable meals. The relationship matrix should be completed through allocation of relative scores to the degree to which each technical descriptor meets each customer requirement.

3-8.   The vision, mission, and strategic plan will be company dependent. Possible vision statement could be, "Become the leader in the logistics industry, internationally." A possible mission statement could be, "Exceed our customer needs, on a global basis, through efficient and effective delivery of goods while catering with a courteous and caring attitude." Strategies may include policies to keep abreast of dynamic customer needs and competitor's performance, adoption of a company culture that promotes innovation through a system of continuous quality improvement, and creating a win-win situation for vendors, customers, employees, and investors.

In a balanced scorecard analysis, in the learning and growth perspective, possible diagnostic measures could be proportion of time problems or delays occur due to failure of information technology (IT) systems, and proportion of failures due to lack of core competencies among staff. Some strategic outcome measures are degree of employee satisfaction, and retention of personnel with core skills. Some strategic performance measures could be the degree of access to strategic information, and amount invested in professional development of technical staff. Under the perspective of internal process, possible diagnostic measures are proportion of deliveries delayed due to lack of facilities (truck, ship, or rail) and proportion of shipments damaged due to mishandling between one form of transport (ship) to another (rail). Some possible strategic outcome measures are operating efficiency of available modes of transportation (ship, rail, truck), and cost per unit volume of shipment in each mode of transportation. Some strategic performance measures could be absenteeism rate of employees directly associated with handling of goods, and degree of investment in new technology.

Under the customer perspective, some diagnostic measures are response time to meet a customer request, and time to process a purchase order and payment for a customer. Some strategic outcome measures are cost of providing a follow-up service, and proportion of satisfied customers. Some performance outcome measures could be the unit cost of subcontracting a segment of the total transportation requirement and the degree of dependability of the subcontractor to contractual obligations. In the financial perspective, some diagnostic measures could be costs incurred for idle storage of goods due to lack of available transport media, and proportion of costs due to breakdown of equipment. Some strategic outcome measures are return on investment, net revenue per unit volume shipped, and total market share. Some possible strategic performance measures are degree of expenses due to rental equipment, and revenue by customer categories based on volume of shipment.

3-9.   The objective statement in the quality function development analysis is to minimize delays in promised delivery dates. Some possible customer requirements and their importance ratings are shown in Table 3-1. While no delay in shipments is the most

TABLE 3-1. Customer Requirements and Importance Ratings

Customer Requirements	Importance Rating
1. No delay in shipments	5
2. Shipment not delivered prior to due date	2
3. Ease of placing order	3
4. On-line tracking capability	4
5. Ease of using credit to place order	3

important (importance rating of 5) requirement, the customer also prefers not to receive the shipment ahead of the promised delivery date, which could be based on following just-in-time (JIT) criteria. Otherwise, there will be a holding or carrying cost of this inventory, if delivered prior to the chosen date. The assigned importance to this requirement is not as much (assigned rating of 2) as that compared to late shipments. Further, ease of placing order is considered moderately important (rating of 3) to the customer. The preference of having an on-line tracking capability, so that the customer may determine the exact location of the shipment in real time, is quite important (with a rating of 4). Additionally, the ease of using customer-available credit to place the order is also of moderate importance.

Some possible means (technical descriptors) to achieve customer requirements are shown in Table 3-2. Six technical descriptors are listed in Table 3-2. Also, for each technical descriptor, the degree of its impact on meeting the five customer requirements is listed. The notation used is as follows: 5 – strong relationship; 3 – medium relationship; 1 – low relationship; and 0 – no relationship. Thus a 1 (5) next to a certain technical descriptor indicates that the technical description has a strong relationship in impacting customer requirement 1, which is no delay in shipments. The remainder of the QFD analysis may be completed through assignment of the appropriate numbers.

3-10. For a company that develops microchips, technological development is one of the major factors that impacts the process of benchmarking. Innovations in chip design are taking place at a rapid level. The steps of benchmarking could be as follows: (1) Decide on measures to benchmark – these could be, for example, physical size of chip, memory capacity of chip, and processing speed; (2) Develop a benchmarking plan – select specific measures on which to focus, identify a process to obtain information from leaders in the field, conduct cost-benefit analysis; (3) Select a method to collect data – type of

TABLE 3-2. Technical Descriptors to Achieve Customer Requirements

Technical Descriptors	Degree of Relationship to Requirements
1. Back-up fleet of carriers through subcontractor	1 (5); 2 (3)
2. Recruit part-time personnel	1 (5); 2 (1)
3. Qualified staff to process order promptly	1 (3); 2 (1); 3 (3)
4. IT software to routinely provide information on product location and the deviation of current time from projected time	1 (5); 2 (5); 3 (3); 4 (5)
5. Web processing using established customer credit and prepare automatic electronic bill	1 (1); 2 (1); 4 (1); 5 (5)

collaboration/relationship necessary to obtain data from other sources, degree of reverse engineering to adopt; (4) Identify sources to benchmark – identify specific sources, some could be available publicly, others could be from competitors that may require mutual agreements; (5) Collect data – obtain data from identified sources; (6) Compare processes – conduct gap analysis between performance of company and that of leader; (7) Make recommendations – determine the type of changes to make in the process to improve microchip performance; (8) Recalibrate benchmarks – make this process on-going.

Top management has the responsibility of ensuring that the design of microchips remains current and competitive. In a field that is constantly evolving, senior management must be knowledgeable of recent developments so that appropriate changes may be made in the process/product design.

3-11. Quality audits are of three types: System audit – this is the most extensive and inclusive type. Here policies, procedures, and operating instructions are evaluated with respect to a reference standard. Further, evaluation of activities and operations to accomplish the desired quality objectives are also conducted. Hence, conformance of quality management standards and their implementation to specified norms are the objectives of such an audit. Such an audit may be used to evaluate a potential vendor. Process audit – this involves an evaluation of selected processes in the organization. These processes are examined and compared to specified standards. While not as extensive as the system audit, such an audit is used to improve processes and have been identified (maybe through Pareto analysis) to be problem areas. Product audit – this involves an assessment of the final product or service to meet or exceed customer requirements. A product audit could determine the effectiveness of a management control system. It is not part of the inspection process. For a company producing multiple products, those that perform poorly could be candidates for such an audit.

3-12. For a financial institution that is considering outsourcing its information technology related services, some criteria to consider are as follows: Error-free performance (reliability) in recording transactions; Ease of access to updated information by identified personnel; Back up procedures to store and retrieve information so that there is no loss of information, potentially; Ease of obtaining summary information as desired by the financial institution (say by type of transaction, account number, etc.); Price of system. The financial institution should next assign weights to the selected criteria, based on its preference. The weights could be on a 100-point scale. Following this, each vendor could be rated on a relative scale (1 to 5), with 1 representing least desirable performance and 5 representing most desirable performance. Finally, a weighted score could be obtained for each vendor, where the weight is multiplied by the rating for each performance measure and then added up. The vendor with the highest weighted score is a candidate for selection.

3-13. Development of drugs for treatment of Alzheimer's disease using nanotechnology is an area of on-going research. Nanotechnology may assist in delivering an appropriate drug to appropriate cells within the human body without causing major side effects. Benchmarking is critical so that time is not wasted in the development process to discover what is already known. Innovation and time-based competition also play an important role since identification of cause-effect relationships and development of appropriate drugs are not necessarily known with certainty. Such processes usually require a good deal of experimentation and testing.

3-14. The mass-transit system in a large city is expected to encounter a projected increase in demand and is considering possible outsourcing.

   a)   A possible mission is: Provide adequate capacity to satisfy projected increase in demand while ensuring customer satisfaction for those who use the mass-transit system. Some objectives could be: Increase capacity at an annual rate that exceeds rate of growth in demand; Ensure travel times meet customer expectations; Reduce delays through efficient scheduling; Provide accident-free services.

   b)   Some criteria for selecting a vendor are as follows: Reliability of operations (as measured by percentage of time trips are on time); Frequency of operations or size of operations that measures number of customers transported between two locations per unit time; Average time to transport customer between two locations by hour of the day – for example, peak hours could be in the morning (7:00 – 8:30 a.m.) and in the afternoon (4:00 – 6:00 p.m.); Price of system.

   c)   Possible diagnostic measures are: Average time to transport passengers between two locations – by hour of the day; Percentage of trips that are late; Total customers transported daily on a weekday; Cost of the system. Some strategic measures are: Percentage of customers satisfied (outcome measure); Return on equity (outcome measure); Percentage utilization of capacity (performance measure).

3-15. There are several benefits of vendor certification. When the vendor is certified such that it consistently meets or exceeds the purchaser's criteria, the need for routine incoming inspection is eliminated. Additionally, a strategic partnership is created between the vendor and the purchaser. Changes in customer needs that require changes in the product, will necessitate changes in raw material or components provided by the vendor. Such customer/product changes are automatically transmitted to the vendors, who make appropriate changes in their processes. Joint effort between the purchaser and vendor helps to reduce lead time and cost, and improve quality.

Typical phases of vendor certification are approved vendor, preferred vendor, and a certified vendor. Initially, the process is documented and performance measures are defined and selected. Roles and responsibilities of the involved personnel are clearly delineated. A quality system survey of the vendor is performed. Vendors that meet acceptable performance standards on defined criteria set by the purchaser are identified as approved vendors.

To move from the approved level to the preferred level, the quality of the vendor's product or service must improve. Hence, the purchaser may assign a preferred status to only the top 10% of its vendors. A vendor with a preferred status may be required to have a process control mechanism in place that shows a focus on problem prevention as opposed to problem detection. A certified vendor, which is the next level, identifies a vendor that not only meets or exceeds the performance measures set by the purchaser, but also has an organizational quality culture that is in consonance with that of the purchaser. In this phase, the vendor and purchaser are partners with a common goal. Both act harmoniously to meet or exceed quality, cost, and delivery goals.

3-16. National and international standards of certification were created to assure uniformity in product/service quality regardless of geographical location of the company. With many companies being multinational in nature, they have plants/branch offices not only in different states/regions of a country but also in several countries. These companies may procure raw material or components from vendors that may be located in several countries. To assure uniformity in the quality of the end product/service, quality issues must be addressed in the vendor companies. This can be accomplished through international certification standards, such as those developed by the International Standards Organization (ISO). One such certification standards is *ISO 9001 – Quality Management Systems – Requirements*. A specific standard for the automobile industry is *ISO/TS 16949*. In the U.S., *QS 9000 – Quality System Requirements*, derived from *ISO 9000* standards, have been adopted by the big three automakers. Thus, for suppliers who are *QS 9000* certified, they may supply to all three automakers without having to face the burden of demonstrating their quality separately to each purchaser.

3-17. Figure 3-1 shows a possible cause-and-effect diagram for lost packages.

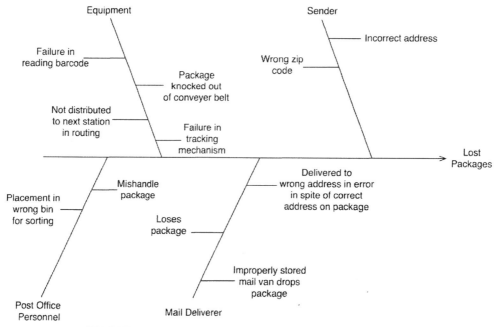

FIGURE 3-1. Cause-and-effect diagram for lost packages

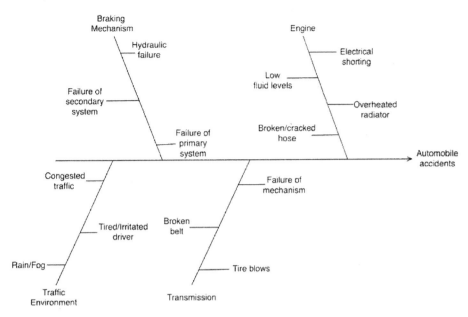

FIGURE 3-2. Cause-and-effect diagram for automobile accidents

3-18. A cause-and-effect diagram for automobile accidents is shown in Figure 3-2.
A FMECA analysis is shown in Table 3-3.

Table 3-3. FMECA for Automobile Accidents

Functional Requirement	Failure Mode	Failure Effects	Severity	Causes	Occurrence	Controls	Detection	RPN
Engine functions appropriately	Engine failure	Severe; Car stalls	9	Radiator malfunction	4	Engine temperature gage	2	72
Engine functions Appropriately	Engine failure	Severe; Car stalls	9	Broken/cracked hose	6	Check hose	5	270
Transmission functions appropriately	Transmission failure	Severe; Car stalls	9	Broken belt	5	Check belt	7	315
Transmission functions appropriately	Tire blows	Car comes to a stop	8	Puncture in tire	4	Check tire	5	160
Primary braking system functions	Primary system fails	Moderate; if secondary system functions	6	Low fluid level	4	Check fluid	6	144
Primary and secondary braking systems function	Brake failure	Severe; Possibility of accident	10	Leaks in fluid vessels	3	Check fluid levels	7	21

Rating scores on severity, occurrence, and detection are assigned and the risk priority number (RPN) is calculated. From the calculated RPN values, the highest value (315) is associated with transmission failure due to broken belts. Some action plans need to be designed to detect such broken or imminent to break belts during routine or preventive maintenance.

3-19. A cause-and effect diagram for automobile accidents caused by the driver is shown in Figure 3-3.

A FMECA analysis is shown in Table 3-4. Rating scores on severity, occurrences, and detection are assigned and the risk priority number (RPN) is calculated. From the calculated RPN values, the highest value (640) is associated with emotionally unfit due to personal issues, followed by emotionally unfit due to disturbed work environment. Detection of such causes are difficult, specially personal issues. This leads to high RPN values which draws attention to create action items to address these issues.

3-20. Depending on the type of product, regions where it is sold, and factors that influence demand, several tools could be used. For example, if there is only one version of the product (with no customer options), we could collect historical data on demand by geographical region (or state), as appropriate. A Pareto chart could depict such demand by region to demonstrate areas of high demand in decreasing order. If we are able to identity factors (or causes) that influence product demand, perhaps a cause-and-effect diagram could be used. Possible causes might be: Population of region, average disposable income, unemployment rate, competitors in the region, number of stores, ease of ordering product, and price. For each cause, certain sub causes could be listed. For example, under competitors in the region, sub causes could be warranty offered, unit price, and lead time to obtain product.

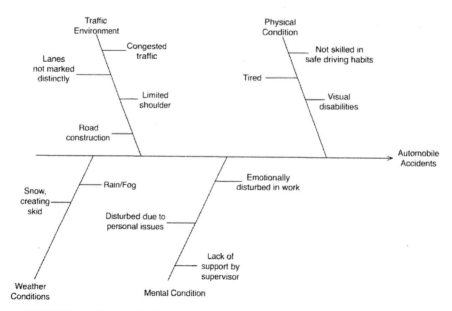

FIGURE 3-3. Cause-and-effect diagram for automobile accidents caused by driver

Table 3-4. FMECA for Automobile Accidents Due to Driver

Functional Requirement	Failure Mode	Failure Effects	Severity	Causes	Occurrence	Controls	Detection	RPN
Physically fit	Physically unfit	Moderate	7	Lack of safe driver skills	4	Study past experience	3	84
Physically fit	Physically unfit	Severe	8	Visual disabilities	7	Have eyes checked	6	336
Physically fit	Physically unfit	Severe	8	Tired	6	Check schedule	8	384
Emotionally fit	Emotionally unfit	Severe	8	Disturbed in work environment	7	Observe behavior	8	448
Emotionally fit	Emotionally unfit	Severe	8	Personal issues	8	Hardly any	10	640
Emotionally fit	Emotionally unfit	Severe	8	Lack of supervisor support	5	Modify supervisor behavior	7	280

3-21. A possible flow chart for visiting the physician's office for a routine procedure is shown in Figure 3-4.

3-22. There are several reasons for failure of total quality management in organizations. First, is the lack of management commitment. While initial enthusiasm may be displayed by management for adoption of TQM, allocation of resources is vital. Release time for managers/staff/employees to devote to quality improvement efforts is critical. These measures may not be implemented by the organization leading to failure in TQM adoption. Second, lack of adoption of a common and consistent company mission that is embraced by all parts of the organization, could be a reason. Often, goals of units within the organization are not coherent with the overall company goals. Decisions made within

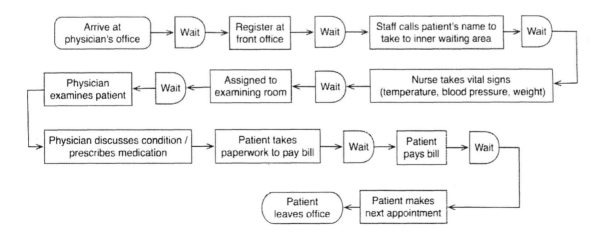

FIGURE 3-4. Flow chart for visit to physician's office

39

a unit are not necessarily optimal from the organizational point of view. Third, lack of efficient flow of information between departments/units could be a factor. Sharing of information may lead to better decisions. Fourth, lack of cross-functional teams to address issues that impact the company could be a reason. Managers may recommend action plans that do not incorporate suggestions from a variety of sub-units that are affected by the decisions.

3-23. Rolled throughput yield $= (0.95)^{20} = 0.3585$.

3-24. Rolled throughout yield now $= (0.98)^{20} = 0.6676$.   So, percentage improvement $= (0.6676 - 0.3585) / 0.3585 = 0.8622 = 86.22\%$.

3-25. Rolled throughout yield with reduced number of operations $= (0.98)^{10} = 0.8171$.   So, percentage improvement $= (0.8171 - 0.6676) / 0.6676 = 0.2239 = 22.39\%$.

3-26. Established standards through such organizations as the *International Standards Organization (ISO)*, *American National Standards Institute (ANSI)*, and *American Society for Quality (ASQ)* serve to create uniformity in audit and certification procedures. Vendors certified through such established standards and using third-party auditors do not need to go through further audits by their customers.

3-27.   a)   A Pareto chart is shown in Figure 3-5.

        b)   The areas to tackle should be inadequate binding, paper tension, and proofreading.

3-28   A scatterplot of life insurance coverage versus disposable income is shown in Figure 3-6.

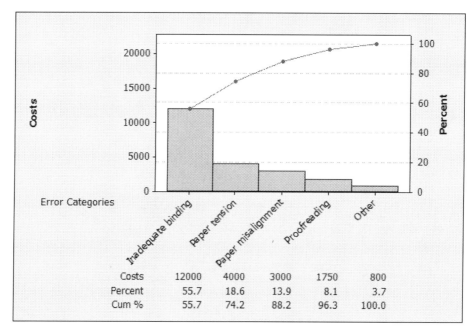

FIGURE 3-5. Pareto chart of error categories

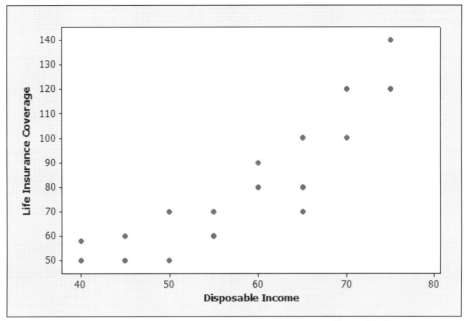

FIGURE 3-6. Scatterplot of life insurance coverage vs disposable income

It seems that, with an increase in disposable income, life insurance coverage increases non-linearly.

3-29.   A flow chart is shown in Figure 3-7.

3-30.   Accomplishing registration to ISO 9001 standards is significantly different from an audit process.   Such registration ensures that an acceptable quality management system is in place.   The system includes processes, products, information systems, documentation, management team, quality culture and traceability, among other items.   When an organization is registered to ISO 9001 standards, it assures customers of a certain level of accomplishment of quality and, therefore, the organization does not necessarily have to

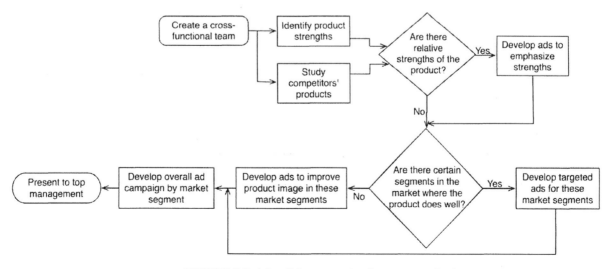

FIGURE 3-7. Advertising campaign for a new product

go through individual audits. Audits, on the other hand, may involve a system audit, process audit, or product audit. They may be internal or external. Audits usually identify deficient areas – they do not necessarily rectify the problem. Development of remedial actions, based on the audit outcomes, is a task for management. Only on implementation of such remedial actions will the benefits be derived.

3-31.   In a global economy, many companies are multinational with branches in several countries. With ISO 9000 standards having a universal impact, registration to such standards creates a seal of acceptance in all of these locations. The customer can trust in the quality management system that exists within the organization, if it is ISO 9000 certified. Hence, the organization does not have to demonstrate its competence by going through other audits. Moreover, if products/components are shipped from a plant in one country to that in another, if they are ISO 9000 certified, the burden of demonstration of the existence of a quality system is reduced. Incoming inspection can be significantly reduced.

3-32.   The Malcolm Baldrige National Quality Award is a national quality award, given annually, in the United States. Three business categories exist – manufacturing, service, and small business. Nonprofit public, private, and government organizations are eligible in a separate category. Also, two other award categories exist – education and health-care. It is not a certification process to standards like ISO 9000 standards.

        The general idea behind the award is to motivate U.S. companies to improve quality and productivity. The objectives are to foster competitiveness. The award winners are expected to share information on their best practices so that other organizations may adopt or benefit from such knowledge. Preparation for filing for the award stimulates a companywide quality effort.

3-33.   a)   The reader may construct a radial plot.

        b)   A matrix plot is shown in Figure 3-8. Based on the matrix plot, to achieve low levels of proportion nonconforming, high levels of temperature, low levels of pressure, high proportion of catalyst and low levels of acidity (pH value) are desirable.

        c)   Figure 3-9 shows a contour plot of proportion nonconforming (varying from levels less than 0.03 to greater than 0.08) for various levels of combinations of pressure and temperature. Similarly, Figure 3-10 shows a contour plot of proportion nonconforming for various levels of combinations of acidity and proportion catalyst.

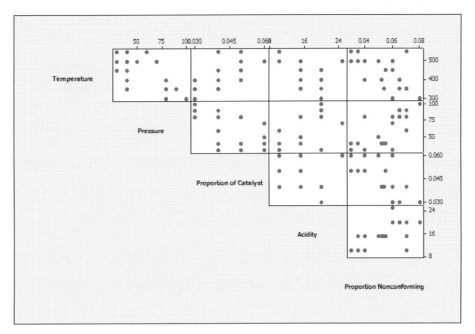

FIGURE 3-8. Matrix plot of temperature, pressure, proportion of catalyst, proportion nonconforming

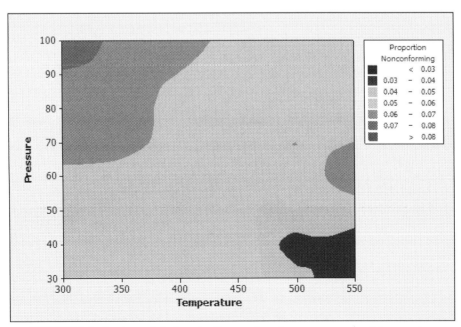

FIGURE 3-9. Contour plot of proportion conforming vs pressure and temperature

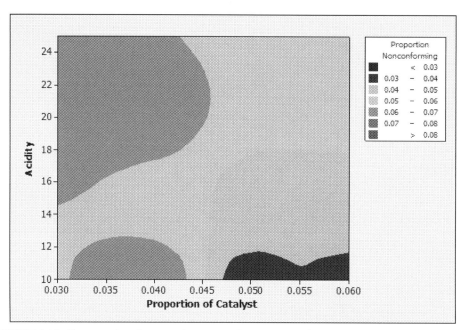

FIGURE 3-10. Contour plot of proportion nonconforming vs. acidity and proportion of catalyst

CHAPTER 4

FUNDAMENTALS OF STATISTICAL CONCEPTS AND TECHNIQUES IN QUALITY CONTROL AND IMPROVEMENT

4-1. a)   Let μ_1 = average life of patients using existing drug; μ_2 = average life of patients using new drug.

$$H_O : \mu_1 - \mu_2 \geq 0; \ H_a : \mu_1 - \mu_2 < 0$$

b)   A type I error occurs when a null hypothesis, that is true, is rejected. So, an existing drug that is at least as good as the new drug, is recommended for replacement with the new drug by the agency. A type II error occurs when a null hypothesis, that is not true, is not rejected. In this situation, the federal agency would not recommend the new drug, even though it increases average life. In the type I error situation, a proven drug would be replaced. It seems that the agency should minimize this risk. In the type II error situation, it would lead to a lost opportunity. Hence, for drug companies, it might add to their research and development expenditures, which may lead to increased unit price of drug to consumers.

c)   Increasing the sample size is a way to reduce the chance of both of these errors.

d)   Normality of distribution of patient life is an assumption. Also, samples are chosen randomly and independently. Depending on assumption on the variability of patient life, for each drug, will influence the precise approach to be used.

4-2. a)   Poisson distribution.

b)   The arrival rate may change in certain 2-hour periods. For example, 11 am to 1 pm may have a higher rate than 8 am to 10 am. In this case, could select only a specified 2-hour period to model. Also, people may flock to store based on crowds, thereby being influenced by others.

c)   Select a given 2-hour period. Observe the number of customers who enter the store during that period. Repeat this for several days. Based on data collected, estimate the mean number of arrivals.

d)   Obtain data on population size, average income, number of similar stores in proximity, etc. for the new location. Based on locations, with similar characteristics, where there are existing stores, estimate the mean number of arrivals in a given time period.

4-3. $P(A \cup B \cup C) = P(A) + P(B) + P(C) - P(AB) - P(BC) - P(AC) + P(ABC)$

4-4.   Accuracy refers to the bias of the measuring instrument. This is the difference between the average of the measured values and the true value. Precision refers to the variation in the measured values. Accuracy is controlled through calibration. Precision is a function of the measuring instrument. Purchasing an instrument with a higher precision is an alternative.

4-5. a)   Ratio

      b)   Ratio

      c)   Ordinal

      d)   Ordinal

      e)   Nominal

      f)   Ordinal

4-6.   The mean is the measure often used in quality control applications. A trimmed mean is preferred when there are outliers in the data that are believed to occur due to special causes.

4-7. a) $H_o: \mu \geq 5$; $H_a: \mu < 5$. Assumptions are normality of distribution of delivery times, and samples are chosen randomly and independently.

      b) $H_o: \mu \geq 10$; $H_a: \mu < 10$. Assumptions are normality of distribution of loan processing times, and samples are chosen randomly and independently.

      c) $H_o: \mu \leq 50,000$; $H_a: \mu > 50,000$. Assumptions are normality of distribution of contract amounts, and samples are chosen randomly and independently.

      d) $H_o: \mu_1 - \mu_2 \leq 0$; $H_a: \mu_1 - \mu_2 > 0$; where μ_1 and μ_2 represent the average response time before and after improvement, respectively. Response times are normally distributed, with random and independent samples chosen. Depending on assumptions on variance of response times, appropriate formula will have to be used.

      e) $H_o: p \leq 0.70$; $H_a: p > 0.70$; where p represents the proportion of customer satisfied with the product. The sample size is large and random samples are chosen.

4-8.   In a binomial distribution, the trials are independent, which they are not in a hypergeometric one. The probability of success on any trial remains constant in a binomial distribution but no so in a hypergeometric one.

4-9.   There is a 95% chance that the mean thickness is contained in the internal. Alternatively, if a large number of such intervals are constructed, about 95% of them will enclose the true mean thickness.

4-10. a) A type I error here implies concluding the mean delivery time is less than 5 days, when in fact it is not. A type II error implies concluding that the mean delivery time is 5 or more days, when in fact it is less. In the first situation, the postal service would be advertising something that they cannot deliver. It may lead to dissatisfied customers. In the second situation, the postal service may miss an opportunity to promote its services. The type I error could be more serious as regards the customer.

b) A type I error implies concluding that the average loan processing time is less than 10 days, when, in fact, it is not. A type II error implies concluding that the average loan processing time is 10 or more days, when in fact it is less. In the first situation; the institution would be raising their customer's expectations, when they may not be able to meet them. It may result in dissatisfied customers. In the second situation, the institution may miss an opportunity to promote itself. The type I error could be more serious as regards the customer.

c) A type I error implies concluding that the average contract amount exceeds $50,000, when in fact it does not. A type II error implies concluding that the average contract amount is no more than $50,000, when in fact it is more. In the first situation, the firm falsely over-projects its customer contracts. If contracts are subject to federal or state restrictions, it could impact them. In the second situation, the firm is under-selling itself. A type I error could be serious under the guidelines of truth-in-advertising. A type II error, in this case, could hurt the firm's chances of obtaining new contracts.

d) A type I error implies concluding that the company has improved its efficiency, when in fact it has not. A type II error implies concluding that the company has not improved its efficiency, when it has. A type I error here could be serious under the guidelines of truth-in-advertising. A type II error here could lead to missed opportunities by failing to publicize its efficient operations.

e) A type I error implies concluding that the proportion of consumers satisfied exceeds 70%, when in fact it does not. A type II error implies concluding that the proportion of satisfied customers does not exceed 70%, when in fact it does. A type I error could be serious in the context of guidelines in truth-in-advertising. A type II error here could lead to missed opportunities.

4-11. The distribution of the price of homes is usually skewed to the right. This is because there are some homes, that are very high-priced, compared to the majority. For such distributions, the median price is a better representative since it is not affected as much as the mean by outliers. The interquartile range will indicate the spread of the prices that are in the middle 50%.

4-12. Let $p \equiv$ proportion that complete the program.

a) $H_o : p \le 0.70$; $H_a : p > 0.70$

b)   A type I error would be committed when we infer that the new program is effective in increasing completion rate when it is really not. In this case, significant costs will be incurred in developing a new program that is not necessarily better than the current one. A type II error would occur when we do not infer that the new program is more effective, when it really is. In this case, we may lose the opportunity of adopting the new program that is more effective.

c)   In addition to identifying a measure of the outcome through the proportion that complete the program and utilizing the cost of development of the new program as a factor for consideration, other factors could be the educational background and level of the persons, age, and number of dependents. Data on the above variables could be collected and a regression or analysis of variance procedure could be used to determine significant factors.

4-13. a)   For patients, let $p \equiv$ proportion that are satisfied. Hypotheses are: $H_o : p \leq p_o, H_a : p > p_o$, where p_o is a specified goal value (say 90%). For employees, a similar set of hypotheses could be tested. For shareholders, a measure of effectiveness could be the unit share price or the rate of return on investment. If μ denotes the average share price or the average rate of return, the hypotheses are: $H_o : \mu \leq \mu_o, H_a : \mu > \mu_o$, where μ_o is a specified goal value.

b)   One measure could be the average waiting time (μ) to see a physician. The hypotheses are: $H_o : \mu \geq \mu_o, H_a : \mu < \mu_o$, where μ_o is a benchmark value.

c)   A measure of the effectiveness of a call center could be the proportion (p) of return calls for the same complaint or information. The hypotheses are: $H_o : p \geq p_o, H_a : p < p_o$, where p_o is a specified value.

d)   Let $\mu \equiv$ average time (or cost) to develop new product. The hypotheses are: $H_o : \mu \geq \mu_o, H_a : \mu < \mu_o$, where μ_o represents a specified value.

4-14. Let $A \equiv \{$product 1 becoming profitable$\}$, $B \equiv \{$product 2 becoming profitable$\}$.

a) $P(AB^c) = P(A) - P(AB) = 0.12 - 0.05 = 0.07$.

b) $P(BA^c) = P(B) - P(BA) = 0.12 - 0.05 = 0.07$.

c) $P(A \cup B) = P(A) + P(B) - P(AB) = 0.12 + 0.12 - 0.05 = 0.19$.

d) $P(A^c \cap B^c) = 1 - P(A \cup B) = 1 - 0.19 = 0.81$.

e) $P(AB^c) + P(BA^c) = P(A \cup B) - P(AB) = 0.19 - 0.05 = 0.14$.

f) $P(B | A) = P(BA)/P(A) = 0.05/0.12 = 0.417$.

4.15.   Let A ≡ {solder defect found}, B ≡ {surface finish defect found}.

a) $P(A \cup B) = P(A) + P(B) - P(AB) = 0.06 + 0.03 - (0.06)(0.03) = 0.0882.$

b) $P(AB^c) = P(A) - P(AB) = 0.06 - 0.0018 = 0.0582.$

c) $P(AB) = P(A)P(B) = (0.06)(0.03) = 0.0018.$

d) $P(A^c \cap B^c) = 1 - P(A \cup B) = 1 - 0.0882 = 0.9118.$

e) $P(A \mid B) = P(AB)/P(B) = 0.0018/0.03 = 0.06.$

4-16.   Let A ≡ {first part defect-free}, B ≡ {second part defect-free}.

a) $P(A) = 1 - 0.05 = 0.95.$

b) $P(B) = 1 - 0.05 = 0.95.$

c) $P(AB) = P(A)\,P(B) = (0.95)(0.95) = 0.9025.$

d) $P(AB^c) + P(BA^c) = (0.95)(0.05) + (0.95)(0.05) = 0.095.$

e) $P(A \cup B) = P(A) + P(B) - P(AB) = 0.95 + 0.95 - 0.9025 = 0.9975.$

4-17.   a)   Sample mean ($\bar{X}$) = 61.4/10 = 6.14. Sample median = 6.2.
Sample mode = 6.2.

b)   Range = 7.9 – 4.5 = 3.4.

Sample variance (s^2) = $\{(5.4 - 6.14)^2 + (6.2 - 6.14)^2 + ... + (6.2 - 6.14)^2\}/9$
$= 11.524/9 = 1.2804.$

Sample standard deviation (s) = $\sqrt{1.2804}$ = 1.1315.

4-18. $P(X = 0) = \dfrac{\binom{5}{0}\binom{20}{4}}{\binom{25}{4}} = 969/2530 = 0.383.$

$P(X = 1) = \dfrac{\binom{5}{1}\binom{20}{3}}{\binom{25}{4}} = 114/253 = 0.4506.$

E(X) = 4(5)/25 = 0.8.

Variance (X) = $\dfrac{4(5)}{25}\left(1 - \dfrac{5}{25}\right)\left(\dfrac{25-4}{25-1}\right)$ = 14/25 = 0.560.

Standard deviation(X) = $\sqrt{0.560}$ = 0.748.

4-19.   n = 5, p = 0.03; P(X = 0) = $\dbinom{5}{0}$ $(0.03)^0(.97)^5 = (.97)^5 = 0.8587$.

P(X = 2) = $\dbinom{5}{2}$ $(0.03)^2(.97)^3 = 10(0.03)^2(0.97)^3 = 0.0082$.

Expected number of nonconforming panels = 1000(0.03) = 30.
Expected cost of rectification = 5(30) = $150.

4-20.   Expected number of computers that will need repair during a year = 20(.08) = 1.6. Without a service contract, expected annual repair costs = $200 (1.6) = $320. If service contract is purchased, annual cost = 20(20) = $400. Thus, based on expected annual costs, university should not buy service contract, since annual expected savings ($400 – $320) of $80 is realized.

Let X be the annual premium per computer for university to be indifferent to buying the service contract. Then annual premium expenses = 20X. We have 20X = 320, which yields X = $16 annual premium per computer.

P(spend no more than $500 annually) $\equiv$ P(X $\leq$ 2), where X represents the number of computers requiring servicing, n = 20, and p = 0.08.

P(X = 0) = $\dbinom{20}{0}$ $(.08)^0 (.92)^{20} = 0.1887$.

P(X = 1) = $\dbinom{20}{1}$ $(.08)^1 (.92)^{19} = 0.3282$.

P(X = 2) = $\dbinom{20}{2}$ $(.08)^2 (.92)^{18} = 0.2711$.

P(X $\leq$ 2) = 0.1887 + 0.3282 + 0.2711 = 0.7880.

4-21.   Use Binomial tables with n = 12, p = 0.10.

a)   P(X $\geq$ 3) = 1 – P(X $\leq$ 2) = 1 – 0.889 = 0.111.

b) $P(X \le 5) = 0.999$.

c) $P(1 \le X \le 5) = P(X \le 5) - P(X \le 0) = 0.999 - 0.282 = 0.717$.

d) Expected number of sensors that will malfunction = $12(0.10) = 1.2$.

e) Standard deviation$(X) = \sqrt{12(0.10)(0.90)} = 1.039$.

4-22. a) Binomial distribution, $n = 40$, $p = 0.05$.

b) $P(X \le 3) = \binom{40}{0} (0.5)^0 (.95)^{40} + \binom{40}{1} (.05)^1 (.95)^{39} + \binom{40}{2} (.05)^2 (.95)^{38}$

$+ \binom{40}{3} (.05)^3 (.95)^{37}$

$= 0.1285 + 0.2705 + 0.2777 + 0.1851 = 0.8618$.

c) Using the Poisson distribution as an approximation to the binomial, $\lambda = np = 40(0.05) = 2$. From cumulative Poisson tables, $P(X \le 3) = 0.857$.

d) There is some discrepancy in the answers between those in parts b) and c). While the binomial distribution is appropriate, the reason for the value obtained using the Poisson distribution to be somewhat less could be that n is not sufficiently large.

4-23. Binomial distribution, $n = 4$, $p = 0.9$. Let $X \equiv$ number of components operating.

P(system functioning) = $P(X \ge 1) = 1 - P(X = 0)$.

$P(X = 0) = \binom{4}{0} (0.9)^0 (0.1)^4 = 0.0001$.

P(system functioning) = $1 - 0.0001 = 0.9999$.
P(system failing) = $P(X = 0) = 0.0001$.

4-24. a) Hypergeometric distribution, $N = 200$, $D = 20$, $n = 10$.
Let $X \equiv$ number of nonconforming fuses in the sample.
Mean = $E(X) = 10(20)/200 = 1$.

$$\text{Var }(X) = \frac{10(20)}{200} \left(1 - \frac{20}{200} \right) \left(\frac{200 - 10}{200 - 1} \right) = 0.859.$$

Standard deviation$(X) = \sqrt{0.859} = 0.927$.

b) Using the binomial distribution as an approximation to the hypergeometric distribution, we have n = 10, p = D/N = 10/200 = 0.1.

$$P(X = 2) = \binom{10}{2} (0.1)^2(0.9)^8 = 0.1937.$$

$P(X \leq 2) = P(X = 0) + P(X = 1) + P(X = 2) = 0.947$ (From cumulative binomial tables).

4-25. a) Poisson distribution, $\lambda = 4$. $P(X = 2) = 0.238 - 0.092 = 0.146$.

b) $P(X \leq 6) = 0.889$.

c) $P(X = 0) = 0.018$.

d) Standard deviation$(X) = \sqrt{\lambda} = \sqrt{4} = 2.0$.

e) From the cumulative Poisson distribution tables, $P(X \leq 7) = 0.949$.
So, if 7 patients are admitted daily, the total expected daily operational expenses is $7(800) = \$5600$.

4-26. a) Poisson distribution, $\lambda = 3$ blemishes/car. $P(X \leq 2) = 0.423$.

b) P(each car has no more than 2 blemishes) = $(0.423)(0.423) = 0.1789$.

c) $\lambda = 6$/two cars. P(no more than 2 blemishes) = 0.062.

4-27. a) Poisson distribution, $\lambda = 7$ failures/year.
$P(X \geq 4) = 1 - P(X \leq 3) = 1 - 0.082 = 0.918$.

b) $P(2 \leq X \leq 8) = P(X \leq 8) - P(X \leq 1) = 0.729 - 0.007 = 0.722$.

c) $\lambda = 14$ failures/two years.
$P(X \leq 8) = 0.062$.

4-28. $\mu = 40$, $\sigma = 2.5$, specification limits are (36, 45).

$$z_1 = \frac{36 - 40}{2.5} = -1.60; \qquad z_2 = \frac{45 - 40}{2.5} = 2.00.$$

Daily cost of scrap = 2000 (0.0548)(0.50) = \$54.80.
Daily cost of rework = 2000 (0.0228)(0.20) = \$9.12.
Total daily cost of rework and scrap = \$63.92.

4-29. $\mu = 4000, \sigma = 25$

$$z = \frac{4050 - 4000}{25} = 2.00.$$

Manufacturer is not meeting the requirement since only 2.28% of the product has a strength that exceeds 4050 kg.

$$z = \frac{x - \mu}{\sigma} \text{ or } -1.645 = \frac{4050 - \mu}{25}$$

$\mu = 4050 + (1.645)(25) = 4091.125$ kg.

4-30. a) $\mu = 0.98, \ \sigma = 0.02$, specification limits are 1.0 ± 0.04 mm.

$$z_1 = \frac{0.96 - 0.98}{0.02} = -1.00$$

$$z_2 = \frac{1.04 - 0.98}{0.02} = 3.00$$

Proportion of conforming washers $= 1 - (0.1587 + 0.0013) = 0.84$.
Daily cost of scrap $= 10000 \ (0.1587)(0.15) = \$238.05$.
Daily cost of rework $= 10000 \ (0.0013)(0.10) = \$ 1.30$.
Total daily cost of rework and scrap $= \$239.35$.

b) $\mu = 1.0, \sigma = 0.02$.

$$z_1 = \frac{0.96 - 1.0}{0.02} = -2.00; \qquad z_2 = \frac{1.04 - 1.0}{0.02} = 2.00.$$

Proportion of rework $= 0.0228$; proportion of scrap $= 0.0228$.
Total daily cost of scrap and rework $= 10000(0.0228)(0.15 + 0.10) = \$57$.

c) $\mu = 1.0, \ \sigma = 0.015$

$$z_1 = \frac{0.96 - 1.0}{0.015} = -2.67; \qquad z_2 = \frac{1.04 - 1.0}{0.015} = 2.67.$$

Proportion of scrap $= 0.0038$; proportion of rework $= 0.0038$. Total daily cost of scrap and rework $= 10000(0.0038)(0.15 + 0.10) = \$9.50$. Percentage decrease in the total daily cost of rework and scrap compared to part a) $= (239.35 - 9.50)/239.35 = 0.9603 = 96.03\%$.

4-31. a) $\mu = 60,000\ kwh,\ \sigma = 400\ kwh.$

$$z = \frac{59,100 - 60,000}{400} = -2.25$$

$P(X < 59,100) = P(Z < -2.25) = 0.0122.$

b) $z_1 = \dfrac{59,000 - 60,000}{400} = -2.50;\qquad z_2 = \dfrac{60,300 - 60,000}{400} = 0.75.$

Probability that monthly consumption will be between 59,000 and 60,300 kwh = $1 - (0.0062 + 0.2266) = 0.7672.$

c) $z = \dfrac{61,100 - 60,000}{400} = 2.75$

$P(\text{Demand less than } 61,100) = P(Z < 2.75) = 0.9970.$

4-32. Exponential time to failure distribution with $\lambda = 1/10,000$.

a) $P(X > 8000) = 1 - P(X \le 8000) = 1 - (1 - e^{-(1/10,000)8000})$
$= 0.449.$

b) $P(X > 15000 \mid X > 9000) = P(X > 6000) = 1 - P(X \le 6000)$
$= 1 - (1 - e^{-(1/10,000)(6000)}) = 0.549.$

c)   Let $A \equiv \{$first component operates for 12,000 hours$\}$,
    $B \equiv \{$second component operates for 12,000 hours$\}$.
$P(A) = 1 - P(X \le 12,000) = 1 - (1 - e^{-(1/10,000)12000}) = 0.30119.$
Similarly, $P(B) = 0.30119.$

$P(A \cup B) = P(A) + P(B) - P(AB)$
$= 0.30119 + 0.30119 - (0.30119)(0.30119)$
$= 0.5117.$

4-33. Time to repair is exponentially distributed with $\lambda = 1/45$.

a) $P(X \le 30) = 1 - e^{-(1/45)30} = 0.4866.$

b) $P(X \le 120) = 1 - e^{-(1/45)120} = 0.9305.$

c)   Standard deviation of $X = 1/\lambda = 45$ minutes.

4-34. a) $\lambda = 1/30;\ P(X > 60) = 1 - P(X \le 60) = 1 - (1 - e^{-(1/30)60})$
$= 0.135.$

b) $P(X \le 65 \mid X > 45) = P(X \le 20) = 1 - e^{-(1/30)20} = 0.4866$.

c)   Let A ≡ {first limousine will not return within 45 min};
            B ≡ {second limousine will not return within 45 min}.

$P(A) = P(X > 45) = 1 - P(X \le 45) = 1 - (1 - e^{-(1/30)45}) = 0.2231$.
Similarly, $P(B) = 0.2231$. P(both not returning within 45 min) = (0.2231)(0.2231)
= 0.0498.

4-35.   Weibull distribution, $\gamma = 0$, ß = 0.25, $\alpha = 800$.

a) $\mu = E(X) = 0 + 800\Gamma(1/0.25 + 1) = 800\Gamma(5) = 19{,}200$ hours

b) $\text{Variance}(X) = 800^2 [\Gamma(2/0.25 + 1) - \{\Gamma(1/0.25 + 1)\}^2]$
            $= 800^2 [\Gamma(9) - \{\Gamma(5)\}^2] = 800^2(39744)$.

Standard deviation$(X) = \sqrt{800^2(39744)} = 159487.178$ hours.

c) $P(X > 1500) = 1 - P(X \le 1500) = 1 - [1 - \exp[-(1500/800)^{0.25}]] = 0.3103$.

4-36.   Weibull distribution, $\gamma = 20$, ß = 0.2, $\alpha = 35$.

a) $P(X \le 30) = 1 - \exp[-\{(30 - 20)/35\}^{0.2}]$
            $= 1 - \exp[-0.77837] = 0.5408$.

b) $E(X) = 20 + 35 \Gamma(1/0.2 + 1) = 20 + 35 \Gamma(6) = 4220$ days.

c) $P(40 \le X \le 50) = P(X \le 50) - P(X \le 40)$
            $= \{1 - \exp[-\{(50-20)/35\}^{0.2}]\} - \{1 - \exp[-\{(40-20)/35\}^{0.2}]\}$
            $= - \exp[-0.9696] + \exp[-0.8941]$
            $= -0.37922 + 0.40897 = 0.02975$.

4-37.   Using the Central Limit Theorem, the sampling distribution of the sample mean ($\overline{X}$) will be approximately normal with mean $\mu_{\bar{x}} = 35$ mm, and standard deviation $\sigma_{\bar{x}} = 0.5/\sqrt{36} = 0.083$.

We want to find: $P(34.95 \le \overline{X} \le 35.18)$.

$z_1 = \dfrac{34.95 - 35}{0.083} = -0.60$; $z_2 = \dfrac{35.18 - 35}{0.083} = 2.1687 \approx 2.17$.

Required probability = 0.9850 − 0.2743 = 0.7107.

4-38. Sampling distribution of $\overline{X}$ is approximately normal with mean $\mu_{\bar{x}} = 35$, and standard deviation $\sigma_{\bar{x}} = 0.5/\sqrt{36} = 0.083$.

a) Want $P(\overline{X} < 34.75) + P(\overline{X} > 35.25)$.

$$z_1 = \frac{34.75 - 35}{0.083} = -3.01, \quad P(\overline{X} < 34.75) = 0.0013.$$

$$z_2 = \frac{35.25 - 35}{0.083} = 3.01, \quad P(\overline{X} > 35.25) = 0.0013.$$

Probability of test concluding that the machine is out of control $= 0.0013 + 0.0013 = 0.0026$.

b) $\mu_{\bar{x}} = 35.05, \quad \sigma_{\bar{x}} = 0.5/\sqrt{36} = 0.083$. Want $P(34.75 \le \overline{X} \le 35.25)$.

$$z_1 = \frac{34.75 - 35.05}{0.083} = -3.614 \approx -3.62.$$

$$z_2 = \frac{35.25 - 35.05}{0.083} = 2.4096 \approx 2.41.$$

Required probability $= 0.9920 - 0.0000 = 0.9920$.

4-39. a) $\mu_{\bar{x}} = 310, \quad \sigma_{\bar{x}} = \sigma/\sqrt{n} = 5/\sqrt{50} = 0.707$. Want $P(\overline{X} \le 308.6)$.

$$z = \frac{308.6 - 310}{0.707} = -1.98.$$

$P(\overline{X} \le 308.6) = P(z \le -1.98) = 0.0239$.

b) $\mu_{\bar{x}} = 310, \quad \sigma_{\bar{x}} = 8/\sqrt{50} = 1.131$, Want $P(\overline{X} \le 308.6)$.

$$z = \frac{308.6 - 310}{1.131} = -1.2378 \approx -1.24.$$

$P(\overline{X} \le 308.6) = P(z \le -1.24) = 0.1075$.

4-40. Exponential distribution with $\lambda = 10^{-3}$ per hour.

a) $E(X) = 1/\lambda = 1000$ hours.

b)   Standard deviation of X = $1/\lambda$ = 1000 hours.

c)   P(X ≥ 1200) = 1 − P (X < 1200).
$$= 1 - [1 - e^{-(1/1000)1200}] = e^{-1.2} = 0.301.$$

d)   P(1200 ≤ X ≤ 1400) = P(X ≤ 1400) − P(X ≤ 1200)

$$= e^{-(1/1000)/1200} - e^{-(1/1000)/1400}$$

$$= e^{-1.2} - e^{-1.4} = 0.054.$$

4-41.   If the time to failure for each switch is distributed exponentially with a failure rate of λ, the number for failures within a certain time t follows a Poisson distribution with parameter λt. Alternatively, time to failure follows a Gamma distribution with k (shape parameter) = 4 and λ = 1/1000.

a)   The switches are identical. With the basic switch and three additional units as standby, the mean time to failure of the system = $(3 + 1)/\lambda$ = $4/10^{-3}$ = 4000 hours.

b)   For a Poisson distribution, the variance equals the mean. So, the standard deviation = $\sqrt{4000}$ = 63.246 hours.

c)   For the system to operate at least 5000 hours, the number of failures must be less than or equal to 3. Using the Poisson distribution, λt = (1/1000)5000 = 5. So, P(X ≤ 3) = 0.265 (using the cumulative Poisson tables).

d)   Let the number of additional switches on a standby basis = a. We may use the Poisson distribution approach (that models number of failures) or the Gamma distribution (that models the time to failure). Using the Poisson distribution, λt = (1/1000)3000 = 3. We desire P(X ≤ a) to be at least 0.40, and determine the minimum value of a that meets this criterion. From the cumulative Poisson distribution tables, P(X ≤ 1) = 0.199 and P(X ≤ 2) = 0.423. So, the minimum number of additional standby switches necessary is 2.

If we use the Gamma distribution, let X denote the time to failure of the system, k = a + 1 (the number of standby units plus the original one), λ =1/1000. We desire P(X ≥ 3000) to be at least 0.40, for the smallest value of a, implying that we want P(X < 3000) to be no more than 0.60. If a = 3, k = 4, using the Gamma distribution, P(X < 3000) = 0.3528. If a = 2, k = 3, P(X < 3000) = 0.5768. For a = 1, k = 2, P(X < 3000) = 0.8008. So, the smallest value of a (the number of standby units) that satisfies the criterion is a = 2.

4-42. a) $E(X) = \exp(\mu + \dfrac{\sigma^2}{2})$

$= \exp(7.6 + 4/2) = \exp(9.6) = 14764.78$ days.

b) $\text{Var}(X) = \exp(2\mu + \sigma^2)[\exp(\sigma^2) - 1]$

$= \exp\{2(7.6) + 4\}[\exp(4) - 1]$

$= \exp(19.2)(53.598) = 1.168 \times 10^{10}$

Standard deviation $(X) = 108,093.933$ days.

c) $P[X > 4000] = P[\ln(x) > 8.294]$

$= 1 - \Phi[(8.294 - 7.6)/2] = 1 - \Phi(0.347) \approx 1 - \Phi(0.35)$

$= 1 - 0.6368 - 0.3632.$

4-43. $n = 40$, $\overline{X} = 10.4$, $\sigma = 1.2$.

a)   90% confidence interval for μ: $\overline{X} \pm z_{\alpha/2}\,\sigma/\sqrt{n}$; $10.4 \pm (1.645)1.2/\sqrt{40} = 10.4 \pm 0.312 = (10.088, 10.712)$. Ninety percent of such constructed intervals will enclose the true average assembly time.

b)   99% confidence interval for μ: $10.4 \pm (2.575)\,1.2/\sqrt{40} = 10.4 \pm 0.489 = (9.911, 10.889)$. Ninety nine percent of such constructed intervals will enclose the true average assembly time.

c)   Samples are chosen randomly and independently and that the population variance is known. The sample size is large enough so that the distribution of the sample mean is normal, using the Central Limit Theorem. However, if the population distribution is normal, the distribution of the sample mean, for any sample size, is also normal.

d) H_0: $\mu \geq 10.8$, H_a: $\mu < 10.8$. Test statistic is given by

$z_0 = \dfrac{\overline{X} - \mu_o}{\sigma/\sqrt{n}} = \dfrac{10.4 - 10.8}{1.2/\sqrt{40}} = -2.108.$

From the standard normal tables, using $\alpha = 0.05$, the critical value of z is -1.645, and the rejection region of the null hypothesis is $z_0 < -1.645$. Since $z_0 = -2.108 < -1.645$, we reject H_0, and conclude that the mean assembly time is less than 10.8 minutes.

4-44. n = 10, $\overline{X}$ = 8.9, s = 0.4737.

a)   95% CI for μ: $8.9 \pm t_{.025,9} (0.4737)/\sqrt{10}$
       $= 8.9 \pm (2.262)(0.1498) = 8.9 \pm 0.3388 = (8.5612, 9.2388).$

Ninety five percent of confidence intervals so constructed will enclose the true mean dissolved oxygen level.

b)   Samples are chosen randomly and independently. The population distribution is assumed to be normal.

c) H_o: $\mu \geq 9.5$, H_a: $\mu < 9.5$. The test statistic is found as

$$t_0 = \frac{\overline{X} - \mu_o}{s/\sqrt{n}} = \frac{8.9 - 9.5}{0.4737/\sqrt{10}} = -4.005.$$

From the t-tables, the critical value of t is $-t_{.05,9} = -1.833$, and the rejection region of the null hypothesis is $t_0 < -1.833$. Since $t_0 = -4.005 < -1.833$, we reject H_o and conclude that the company is violating the standard.

4-45. Let $\mu_1 \equiv$ mean time lost before implementation of OSHA program, and $\mu_2 \equiv$ mean time lost after implementation of OSHA program. It is given that $\sigma_1 = \sigma_2 = 3.5$ hours, $n_1 = 40$, $\overline{X}_1 = 45$, $n_2 = 45$, and $\overline{X}_2 = 39$.

a)   90% CI for $(\mu_1 - \mu_2)$: $(45-39) \pm z_{.05} \sqrt{\dfrac{3.5^2}{40} + \dfrac{3.5^2}{45}}$

$= 6 \pm (1.645)(0.7606) = 6 \pm 1.251 = (4.749, 7.251).$

b) H_o: $\mu_1 - \mu_2 \leq 0$, H_a: $\mu_1 - \mu_2 > 0$. The test statistic is

$$z_0 = \frac{(45-39)-0}{\sqrt{\dfrac{3.5^2}{40} + \dfrac{3.5^2}{45}}} = 7.889.$$

The critical value of z is 1.282, with the rejection region of the null hypothesis being $z_0 > 1.282$. Since $z_0 = 7.889 > 1.282$, we reject H_o and conclude that implementation of the OSHA program has reduced the mean employee lost time.

4-46. The population standard deviations σ_1 and σ_2 are unknown but assumed to be equal. We have $n_1 = 40$, $\overline{X}_1 = 45$, $s_1 = 3.8$, $n_2 = 45$, $\overline{X}_2 = 39$, $s_2 = 3.5$.

a)   95% CI for ($\mu_1 - \mu_2$): $(45-39) \pm t_{.025,83}\, s_p \sqrt{\dfrac{1}{40} + \dfrac{1}{45}}$.

From the t-tables, $t_{.025,83} \approx 1.989$. The pooled estimate of the common variance is

$$s_p^2 = \frac{39(3.8)^2 + 44(3.5)^2}{83} = 13.881; \; s_p = \sqrt{13.881} = 3.726.$$

95% CI for ($\mu_1 - \mu_2$): $6 \pm (1.989)(3.726)\sqrt{\dfrac{1}{40} + \dfrac{1}{45}}$

$= 6 \pm 1.6105 = (4.3895, 7.6104)$.

b)   Samples are randomly chosen and that each population has a normal distribution.

c) H_0: $\mu_1 - \mu_2 \le 0$, H_a: $\mu_1 - \mu_2 > 0$. The test statistic is

$$t_0 = \frac{(45 - 39) - 0}{3.726\sqrt{\dfrac{1}{40} + \dfrac{1}{45}}} = 7.410.$$

The critical value of t is $t_{.05,83} \approx 1.6634$, with the rejection region of the null hypothesis being $t_0 > 1.6634$. Since $t_0 = 7.410 > 1.6634$, we reject H_0 and conclude that the mean employee lost time has decreased due to the OSHA program.

d) H_0: $\sigma_1^2 = \sigma_2^2$, H_a: $\sigma_1^2 \ne \sigma_2^2$. The test statistic is

$$F_0 = s_1^2/s_2^2 = (3.8)^2/(3.5)^2 = 1.1788.$$

The critical values are $F_{.025,39,44}$ and $F_{.975,39,44}$, with the rejection region of the null hypothesis being $F_0 > F_{.025,39,44} \approx 1.8607$ (on interpolation) or $F_0 < F_{.975,39,44}$. The test statistic ($F_0 = 1.1788$) does not fall in the rejection region of the null hypothesis, so we do not reject H_0. Hence we have not rejected the hypothesis of equal variances of the two populations.

4-47. Let $\sigma_1^2 \equiv$ variance of breaking strength of natural fiber,

$\sigma_2^2 \equiv$ variance of breaking strength of synthetic fiber.

We are given that $n_1 = 8$, $\overline{X}_1 = 540$, $s_1 = 55$, $n_2 = 10$, $\overline{X}_2 = 610$, $s_2 = 22$.

a) H_0: $\sigma_1^2 = \sigma_2^2$, H_a: $\sigma_1^2 \ne \sigma_2^2$. The test statistic is

$F_o = s_1^2 / s_2^2 = (55)^2/(22)^2 = 6.25.$

The critical values are $F_{.025,7,9}$ and $F_{.975,7,9}$. From the F-tables, $F_{.025,7,9} = 4.20$. Now, $F_{.975,7,9} = 1/F_{.025,9,7} = 1/4.82 = 0.207$. The rejection region of the null hypothesis is $F_o > 4.20$ or $F_o < 0.207$. Since the test statistic, $F_o = 6.25 > 4.20$, we reject H_o and conclude that the population variances are not equal. The assumptions necessary to perform this test are that the samples are random and independent and that each population has a normal distribution.

b) $H_o: \mu_1 - \mu_2 \geq 0, \quad H_a: \mu_1 - \mu_2 < 0.$ The test statistic is

$$t_o = \frac{(540 - 610) - 0}{\sqrt{\dfrac{(55)^2}{8} + \dfrac{(22)^2}{10}}} = -3.389.$$

The degrees of freedom of t is:

$$\nu = \frac{\left(\dfrac{55^2}{8} + \dfrac{22^2}{10}\right)^2}{\dfrac{(55^2/8)^2}{7} + \dfrac{(22^2/10)^2}{9}}$$

$$= 8.778.$$

The critical value of t is $-t_{.10,8.778} \approx -1.386$, with the rejection region of the null hypothesis being $t_o < -1.386$. Since $t_o = -3.389 < -1.386$, we reject H_o and conclude that the mean breaking strength of synthetic fibers exceeds that of natural fibers.

c)   95% confidence interval for σ_1^2/σ_2^2 is:

$$\frac{s_1^2}{s_2^2} \frac{1}{F_{.025,\nu_1,\nu_2}} \leq \frac{\sigma_1^2}{\sigma_2^2} \leq \frac{s_1^2}{s_2^2} F_{.025,\nu_1,\nu_2}$$

$$\frac{(55)^2}{(22)^2} \frac{1}{F_{.025,7,9}} \leq \frac{\sigma_1^2}{\sigma_2^2} \leq \frac{(55)^2}{(22)^2} F_{.025,9,7}$$

$$\frac{(55)^2}{(22)^2} \frac{1}{4.20} \leq \frac{\sigma_1^2}{\sigma_2^2} \leq \frac{(55)^2}{(22)^2} 4.82$$

$$1.489 \leq \frac{\sigma_1^2}{\sigma_2^2} \leq 30.125.$$

d) 90% CI for $(\mu_2 - \mu_1)$:

$$(610 - 540) \pm t_{.05,8.778} \sqrt{\frac{(22)^2}{10} + \frac{(55)^2}{8}}$$

$$70 \pm (1.839) \sqrt{426.525} = 70 \pm 37.98 = (32.02, 107.98).$$

4-48. Sample mean and standard deviation are given by $\overline{X} = 6.14$, and s = 1.1316, respectively.

a) 98% CI for μ: $6.14 \pm t_{.01,9} (1.1316)/\sqrt{10} = 6.14 \pm (2.821)(0.3578)$
    $= 6.14 \pm 1.009 = (5.131, 7.149).$

Samples are random and the population distribution is normal.

b) 95% CI for σ^2:

$$\frac{9(1.1316)^2}{\chi_{.025,9}^2} \leq \sigma^2 \leq \frac{9(1.1316)^2}{\chi_{.975,9}^2}$$

$$\frac{9(1.1316)^2}{19.02} \leq \sigma^2 \leq \frac{9(1.1316)^2}{2.70} \text{ , or}$$

$$0.6059 \leq \sigma^2 \leq 4.2684$$

c) H_o: $\sigma^2 \leq 0.80$, H_a: $\sigma^2 > 0.80$. The test statistic is

$$\chi_0^2 = \frac{9(1.1316)^2}{0.80} = 14.406.$$

The critical value is $\chi_{.05,9}^2 = 16.92$, with the rejection region given by $\chi_0^2 > 16.92$. Since $\chi_0^2 = 14.406 < 16.92$, we do not reject H_o. So, we cannot conclude that the process variance exceeds 0.80.

4-49. a) Point estimate $= \hat{p} = 80/300 = 0.267.$

b) 95% CI for p: $0.267 \pm z_{.025} \sqrt{\frac{(0.267)(0.733)}{300}}$

$$= 0.267 \pm 1.96\ (0.02554) = 0.267 \pm 0.050 = (0.217, 0.317).$$

c) H_o: $p \le 0.25$, H_a: $p > 0.25$. The test statistic is:

$$z_o = \frac{0.267 - 0.25}{\sqrt{\dfrac{(0.25)(0.75)}{300}}} = 0.68$$

The critical value is $z_{.01} = 2.33$, with the rejection region being $z_o > 2.33$. Since the test statistic (z_o) = 0.68 < 2.33, we do not reject H_o, and so we cannot claim that the market share is more than 25%.

4-50.   Let $p_1 \equiv$ proportion of people who prefer the product before the advertising campaign, $p_2 \equiv$ proportion of people who prefer the product after the advertising campaign.

$\hat{p}_1 = 40/200 = 0.2$, $\hat{p}_2 = 80/300 = 0.267$, $n_1 = 200$, $n_2 = 300$.

a)   90% CI for ($p_1 - p_2$): $(0.2 - 0.267) \pm z_{.05}\ \sqrt{\dfrac{(0.2)(0.8)}{200} + \dfrac{(0.276)(0.733)}{300}}$

$$= -0.067 \pm 1.645\ (0.0381) = -0.067 \pm 0.0627 = (-0.1297, -0.0043).$$

b) H_o: $p_1 - p_2 \ge 0$, H_a: $p_1 - p_2 < 0$. The pooled estimate is:

$$\hat{p} = \frac{200(40/200) + 300(80/300)}{200 + 300} = \frac{40 + 80}{500} = 0.24.$$

The test statistic is:

$$z_o = \frac{(0.2 - 0.267)}{\sqrt{(0.24)(0.76)\left(\dfrac{1}{200} + \dfrac{1}{300}\right)}} = -1.7185.$$

The critical value is $-z_{.10} = -1.282$, with the rejection region being $z_o < -1.282$. The test statistic of $-1.7185 < -1.282$, so we reject H_o and conclude that the advertising campaign has been successful in increasing the proportion of people who prefer the product.

4-51. $\hat{p}_1 = 6/80 = 0.075$, $\hat{p}_2 = 14/120 = 0.117$, $n_1 = 80$, $n_2 = 120$

a) H_o: $p_1 - p_2 = 0$, H_a: $p_1 - p_2 \ne 0$.

The pooled estimate is $\hat{p} = \dfrac{6 + 14}{20 + 120} = 0.10$.

The test statistic is:

$$z_0 = \frac{(0.75 - 0.117)}{\sqrt{(0.10)(0.90)\left(\dfrac{1}{80} + \dfrac{1}{120}\right)}} = -0.970$$

The critical values are $\pm z_{.05} = \pm 1.645$, with the rejection region of the null hypothesis being $z_0 < -1.645$ and $z_0 > 1.645$. Since the test statistic is -0.970, which does not lie in the rejection region, we do not reject H_o. We conclude that there is no difference in the output of the machines as regards the proportion of nonconforming parts.

b)   95% CI for $(p_1 - p_2)$:

$$(0.075 - 0.117) \pm z_{.025} \sqrt{\frac{(0.075)(0.925)}{80} + \frac{(0.117)(0.883)}{120}}$$

$$= -0.042 \pm 1.96(0.04157) = -0.042 \pm 0.0815 = (-0.1235, 0.0395).$$

4-52. $n = 10$, $\overline{X} = 9.91$, $s = 0.2767$, $s^2 = 0.0765$.

a)   90% CI for σ^2:

$$\frac{9(0.0765)}{\chi^2_{.05,9}} \leq \sigma^2 \leq \frac{9(0.0765)}{\chi^2_{.95,9}} ,$$

$$\frac{9(0.0765)}{16.92} \leq \sigma^2 \leq \frac{9(0.0765)}{3.33}$$

$$0.0407 \leq \sigma^2 \leq 0.2067 .$$

b)   90% CI for σ:

$$\sqrt{0.0407} \leq \sigma \leq \sqrt{0.2067}$$

$$0.2017 \leq \sigma \leq 0.4546 .$$

c) H_o: $\sigma^2 \leq 0.05$, H_a: $\sigma^2 > 0.05$. The test statistic is

$$\chi^2_0 = \frac{9(0.0765)}{0.05} = 13.77.$$

The critical value is $\chi^2_{.10,9} = 14.68$, with the rejection region of the null hypothesis being $\chi^2_0 > 14.68$. Since the test statistic of $13.77 < 14.68$, we do not reject H_o. We conclude that the variance of the diameters does not exceed 0.05.

d) H_o: $\mu = 9.5$, H_a: $\mu \neq 9.5$. The test statistic is:

$$t_o = \frac{9.91 - 9.5}{0.2767/\sqrt{10}} = 4.686.$$

The critical value is $\pm t_{.05,9} = \pm 1.833$, with the rejection region being $t_o < -1.833$ and $t_o > 1.833$. Since the test statistic of $4.686 > 1.833$, we reject H_o and conclude that the mean differs from 9.5.

4-53. $n_1 = 10$, $\overline{X}_1 = 4.5$, $s_1 = 2.3$, $n_2 = 12$, $\overline{X}_2 = 3.4$, $s_2 = 6.2$.

a) 90% CI for σ_1^2/σ_2^2:

$$\frac{(2.3)^2}{(6.2)^2} \frac{1}{F_{.05,9,11}} \leq \frac{\sigma_1^2}{\sigma_2^2} \leq \frac{(2.3)^2}{(6.2)^2} F_{.05,11,9} \quad,$$

$$\frac{(2.3)^2}{(6.2)^2} \frac{1}{2.90} \leq \frac{\sigma_1^2}{\sigma_2^2} \leq \frac{(2.3)^2}{(6.2)^2} 3.114 \quad,$$

$$0.0474 \leq \frac{\sigma_1^2}{\sigma_2^2} \leq 0.4285 .$$

b) Each population is normally distributed. Random samples chosen from each population.

c) H_o: $\sigma_1^2 \geq \sigma_2^2$, H_a: $\sigma_1^2 < \sigma_2^2$, or written alternatively as:

H_o: $\sigma_2^2 \leq \sigma_1^2$, H_a: $\sigma_2^2 > \sigma_1^2$. The test statistic is

$F_o = s_2^2/s_1^2 = (6.2)^2/(2.3)^2 = 7.266$.

The critical value is $F_{.05,11,9} = 3.114$, with the rejection region being $F_o > 3.114$. Since the test statistic of $7.266 > 3.114$, we reject H_o and conclude that the first vendor has smaller variability than that of the second.

If we had chosen to write the hypothesis as: H_o: $\sigma_1^2 \geq \sigma_2^2$, H_a: $\sigma_1^2 < \sigma_2^2$, and constructed the test statistic as:

$F_o = \dfrac{s_1^2}{s_2^2} = (2.3)^2/(6.2)^2 = 0.1376$, the critical value would be

$F_{.95,9,11} = \dfrac{1}{F_{.05,11,9}} = \dfrac{1}{3.114} = 0.3211$, with the rejection region being

$F_o < 0.3211$. Since the test statistic of $0.1376 < 0.3211$, we reject H_o and conclude that the first vendor has smaller variability than that of the second.

d) Since the first vendor has a smaller variability of the delay time, compared to that of the second, it would be more reliable. However, because the sample average delay time of the first (4.5 days) is greater than that of the second (3.4 days), we need to test if the population mean delay time of the first vendor exceeds that of the second.

The hypotheses to be tested are: H_o: $\mu_1 - \mu_2 \le 0$, H_a: $\mu_1 - \mu_2 > 0$. The test statistic is:

$$t_o = \dfrac{(4.5 - 3.4)}{\sqrt{\dfrac{2.3^2}{10} + \dfrac{6.2^2}{12}}} = 0.569.$$

The degrees of freedom of the test statistic is:

$$v = \dfrac{\left(\dfrac{2.3^2}{10} + \dfrac{6.2^2}{12}\right)^2}{\dfrac{(2.3^2/10)^2}{9} + \dfrac{(6.2^2/12)^2}{11}}$$

$$= 14.45.$$

For a chosen level of significance (α) of .05, the critical value is $t_{.05,14.45} \approx 1.757$, with the rejection region being $t_o > 1.757$. Since the test statistic of $0.569 < 1.757$, we do not reject H_o. Thus, we cannot conclude that the mean delay of the first vendor exceeds that of the second.

So, on the basis of the test on comparison of variances of delay time, one would select the first vendor.

4-54. a) Minitab was used to obtain the following results. Systolic blood pressure (before administration of drug): Mean = 133.04; Standard deviation = 14.56; Skewness coefficient = 0.03; Kurtosis coefficient = – 0.65; IQR = 20.50. So distribution is nearly symmetric about mean and less peaked than the normal.

b) For systolic blood pressure after administration of drug: Mean = 125.80; Standard deviation = 9.39: Skewness coefficient = 0.33; Kurtosis coefficient = -0.54; IQR= 13.00. The mean and standard deviation are both lower than their corresponding values before administration of the drug. We will test, later, if there has been a significant decrease in the mean value.

c) $H_o : \mu \leq 125$; $H_a : \mu > 125$

Test statistic = $t_o = \dfrac{133.04 - 125}{14.56 / \sqrt{25}} = 2.76.$

The critical value of $t_{.05,24} = 1.711$, with rejection region of H_o being $t_o > t_{.05,24}$. Since $t_o = 2.76 > 1.711$, we reject H_o and conclude that the mean systolic blood pressure, before administration of the drug, exceeds 125. Incidentally, the p–value = $P[t > 2.76] = .005 < \alpha = 0.05$.

d)   Paired difference t-test with $H_o : \mu_1 - \mu_2 \leq 0$ vs. $H_a : \mu_1 - \mu_2 > 0$, where μ_1 and μ_2 represent the mean systolic blood pressure before and after administration of the drug, respectively.

Test statistic = $t_o = \dfrac{7.24 - 0}{7.18 / \sqrt{25}} = 5.04.$

Probability value (p–value) = $0.000 < \alpha = .05$, so we reject H_o and conclude that the drug was effective in reducing mean systolic blood pressure. The p–value states that, if the null is true (i.e., no impact of the drug), the chance of observing an average difference of 7.24 or even more extreme, is extremely small (0.000). Hence, we reject H_o.

e)   Paired difference t–test on average cholesterol values with $H_o : \mu_1 - \mu_2 \leq 0$ vs. $H_a : \mu_1 - \mu_2 > 0$, where μ_1 and μ_2 represent the average cholesterol level before and after administration of the drug, respectively.

Test statistic = $t_o = \dfrac{5.32 - 0}{10.61 / \sqrt{25}} = 2.51.$

The p–value = $0.010 < \alpha = 0.05$. So, we reject H_o and conclude that the drug was effective in reducing average cholesterol level.

f)   Descriptive statistics on blood glucose level before administration of the drug: Mean = 138.52; Standard deviation = 31.23; Skewness coefficient = 1.38; Kurtosis coefficient = 0.98; IQR = 28.50. The distribution is skewed to the right and somewhat more peaked than the normal.

g) The correlation coefficient (r) between blood glucose levels before and after administration of the drug = 0.959. Testing, $H_o : \rho = 0$ vs. $H_a : \rho \neq 0$, yields a test statistic of:

$$t_o = \frac{0.959\sqrt{23}}{\sqrt{1-0.959^2}} = 16.228.$$

Since $t_o > t_{.025,23} = 2.069$, we reject H_o and conclude that the correlation coefficient differs from zero.

h) A 98% confidence interval for the variance of systolic blood pressure after administration of the drug is given by:

$$\frac{24(9.39^2)}{\chi^2_{.01,24}} \leq \sigma^2 \leq \frac{24(9.39)^2}{\chi^2_{.99,24}}$$

$$\frac{24(9.39)^2}{42.98} \leq \sigma^2 \leq \frac{24(9.39)^2}{10.86}$$

$$49.235 \leq \sigma^2 \leq 194.855.$$

4.55. a) Minitab was used to obtain the following results. Processing time prior to changes: Mean = 10.165; Standard deviation = 1.174; Skewness coefficient = 0.04; Kurtosis coefficient = -1.01; IQR = 2.100. Distribution is fairly symmetric about mean but flatter than the normal distribution.

b) Processing time after changes: Mean = 8.077; Standard deviation = 0.869; Skewness coefficient = -0.11; Kurtosis coefficient = -0.37; IQR = 1.175. Distribution is skewed to the left and flatter than the normal distribution.

c) 95% confidence internal for mean processing time prior to changes:

$$10.165 \pm t_{.025,30} (1.174)/\sqrt{31}$$

$$10.165 \pm 2.042(1.174)/\sqrt{31}$$

$$10.165 \pm 0.431 = (9.734, 10.596).$$

d) $H_o : \mu \geq 10.5$ vs. $H_a : \mu < 10.5$. The test statistic is:

$$t_o = \frac{10.165 - 10.5}{1.174 / \sqrt{31}} = -1.589.$$

The rejection region of H_o is $t_o < t_{.02,30}$. Here, the p–value is $P(t \le -1.589) = 0.061$. Since p–value $= 0.061 > \alpha = 0.02$, we do not reject H_o. If $\alpha = 0.10$, the p–value would be less than α, and we would reject H_o. Hence, the decision is dependent on the chosen level of significance (α).

e) $H_o : \mu \ge 8.5;\ H_a : \mu < 8.5$. The test statistic is:

$$t_o = \frac{8.077 - 8.5}{0.869 / \sqrt{26}} = -2.48$$

The rejection region of H_o is $t_o < -t_{.05,25} = -1.708$. Since the test statistic of $-2.48 < -1.708$, we reject H_o and conclude that the mean processing time after changes is less than 8.5 days. The p-value $= 0.010$. This means, if the null hypothesis is true, the chances of getting a sample average of 8.077 or less is very small (only 0.010). So, we reject the null hypothesis since the p-value $< \alpha = 0.05$.

f) $H_o :\ \sigma_1^2 = \sigma_2^2;\ H_a :\ \sigma_1^2 \ne \sigma_2^2$, where σ_1^2 and σ_2^2 represent the variance of the processing time before and after process changes, respectively. Test statistic is:

$$F = \frac{s_1^2}{s_2^2} = \frac{(1.1743)^2}{(0.8687)^2} = 1.827.$$

The critical values are: $F_{.025,30,25} = 2.18$ and $\dfrac{1}{F_{.025,25,30}} = 1/2.12 = 0.4716$.

Since the test statistic falls between the critical values, we do not reject the null hypothesis. Using the p-value approach, p-value $= 0.128 > \alpha = 0.05$.

g) $H_o : \mu_1 - \mu_2 \le 0;\ H_a : \mu_1 - \mu_2 > 0$. In part f), since we did not reject the null hypothesis of equality of variances, we use the t-test that makes use of the pooled estimate of the sample variances:

$$s_p^2 = \frac{30(1.174)^2 + 25(0.869)^2}{55} = 1.095,\ s_p = 1.0465.$$

The test statistic is:

$$t_o = \frac{10.165 - 8.077}{1.0465\sqrt{\dfrac{1}{31} + \dfrac{1}{26}}} = 7.50.$$

The rejection region is $t_o :> t_{.05,55} = 1.673$. Since the test statistic $= 7.50 >$ 1.673, we reject H_o and conclude that the process changes have been effective in reducing mean processing time. Further, the p-value $= 0.000$. The assumptions made are that the distribution of processing times are normal, for both before and after process changes. Also, random and independent samples are chosen from each process.

4-56. Let the premium to be charged be denoted by p. The probability distribution of X, the net amount retained by the company is given as follows:

X	$P(X = x)$
p	0.9885
$p - 200,000$	0.0005
$p - 100,000$	0.001
$p - 50,000$	0.01

Since the company wants to make an average net profit of 1.5% of the face value of the policy, we have:

$E(X) = 3000$, or

$$p(0.9885) + (p - 200,000)(0.0005) + (p - 100,000)(0.001) + (p - 50,000)(0.01) = 3000$$

or $p = \$3700$.

4-57. a) Average number of errors in 2 hours of operation $= (300 \times 5000 \times 2)/10^6 = 3 = \lambda$. Assuming a Poisson distribution of occurrence of errors. $P(X \le 5) = 0.916$ (From the cumulative Poisson distribution).

b) For two hours of operation, $\lambda = 3$. $P(X = 0) = 0.50$.

For three hours of operation, $\lambda = (300 \times 5000 \times 3)/10^6 = 4.5$. $P(X = 0) = 0.011$.

c) Desire $P(X = 0) = 0.001$. Let λ denote the hourly error rate. So, for 2 hours of operation, average $= 2\lambda$.

We have $P(X = 0) = e^{-2\lambda} = 0.001$. Taking natural logarithm yields, $-2\lambda = -6.9077$, or $\lambda = 3.45388$ per hour.

4-58. a) Assuming a Poisson distribution of the occurrence of errors, the average number of errors per circuit board = λ = (200 x 5000)/10^6 = 1. $P(X \geq 3)$ = 1 – $P(X \leq 2) = 0.920$.

b) $P(X = 0) = 0.368$.

c) Average number of errors per new circuit board = λ = (200 x 2000)/10^6 = 0.4. $P(X = 0) = 0.670$.

d) Expected cost reduction per month = $(1 – 0.4)10^6$ x 0.05 = $30,000.

CHAPTER 5

DATA ANALYSES AND SAMPLING

5-1. Confidence interval or hypothesis testing on the population mean when the population standard deviation is not known. Here, the t-statistic is used and it requires the distributional assumption of normality. Other parametric tests may be hypothesis testing on the difference in the means of two populations, when the population variances are not known, hypothesis testing on the population variances, or that on comparing two population variances. All of these tests require the assumption of normality of the distribution of the characteristic. If the assumption is not satisfied, transformations may be considered. These include power transformations or Johnson's transformation.

5-2. Stratified random sampling with a proportional allocation scheme.

5-3. Chi-squared test for independence of classifications in contingency tables.

5-4. Chi-squared test on cell probabilities.

5-5. The various parameters and sample size are related. For example, for a given type I error and power, as the degree of difference that one wishes to detect in a parameter decreases, the sample size increases and vice versa. For a given difference in the parameter, the type I error can be reduced and the power increased by increasing the sample size.

5-6. H_o: No billing errors; H_a: Billing errors. A type I error implies concluding that there are billing errors in a customer account when there are none. This could result in a wasted effort and cost on the part of auditors to detect billing errors. A type II error implies concluding that there are no billing errors when, in fact, they exist. Here, customers who find errors in their bills could be very dissatisfied, leading to loss of future market share.

5-7. A stratified random sampling procedure using a proportional allocation scheme could be chosen. The strata would be defined by the different time periods in which the traffic rate varies.

5-8. a) A histogram is shown in Figure 5-1. Mean = 30.782; Median = 30.4; Standard deviation = 2.7648. Distribution is skewed to the right.

   b) $H_o : \mu \geq 32$; $H_a : \mu < 32$. Assumptions necessary are that the distribution of waiting time is normal, and random, independent samples are selected.

   c) A transformation of the type natural logarithm of the waiting time is explored, given the positively skewed distribution of waiting times. A normality test using the Anderson-Darling test in Minitab is conducted on the natural logarithm of waiting time. The p-value is 0.117, thus validating the normality of the transformed variable (see Figure 5-2).

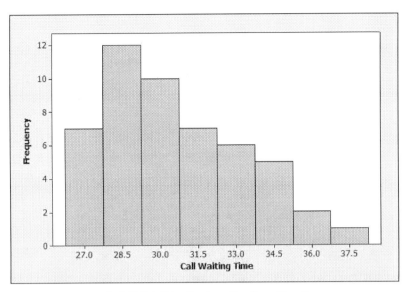

FIGURE 5-1. Distribution of call waiting time

d)   Testing $H_o : \mu \geq 32$ vs. $H_a : \mu < 32$, where μ represents the mean waiting time in seconds is equivalent to testing hypothesis on the transformed variable: $H_o : \mu' \geq 3.4657$ vs. $H_a : \mu' < 3.4657$, where μ' represents the natural logarithm of the waiting time in seconds. The mean and standard deviation of the transformed variable are 3.423 and 0.888, respectively. The test statistic is:

$$t_o = \frac{3.423 - 3.4657}{0.088 / \sqrt{50}} = -3.43$$

The p-value is $0.001 < \alpha = .05$, so we reject the null hypothesis and conclude the mean waiting time is less than 32 seconds.

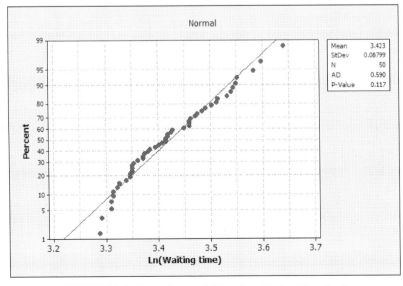

FIGURE 5-2. Normal probability plot of ln(waiting time)

```
Stem-and-leaf of Call Waiting Time  N  = 50
Leaf Unit = 0.10

         2    26   89
         8    27   445578
        16    28   24455568
        22    29   112458
        (8)   30   02445578
        20    31   58889
        15    32   2368
        11    33   256
         8    34   25689
         3    35
         3    36   05
         1    37
         1    38   1
```

FIGURE 5-3. A stem and leaf plot for call waiting time

e)   To find a 90% confidence interval for the variance, let us first find the confidence interval for the transformed variable (natural logarithm of waiting time).

$$\frac{49(3.423)^2}{66.3386} \leq \sigma'^2 \leq \frac{49(3.423)^2}{33.9303}, \text{ or}$$

$$8.6545 \leq \sigma'^2 \leq 16.9208.$$

5-9.   A stem-and-leaf plot is shown in Figure 5-3. The distribution is skewed to the right. A box plot is shown in Figure 5-4. Note the long top whisker and the relatively short bottom whisker indicating a right-tailed distribution. A 95% confidence interval for the median is given by:

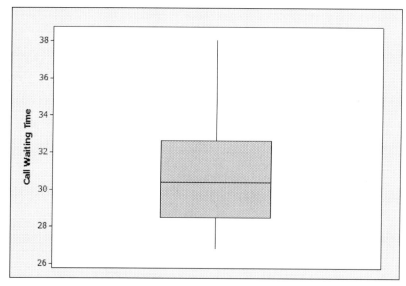

FIGURE 5-4. Boxplot of call waiting time

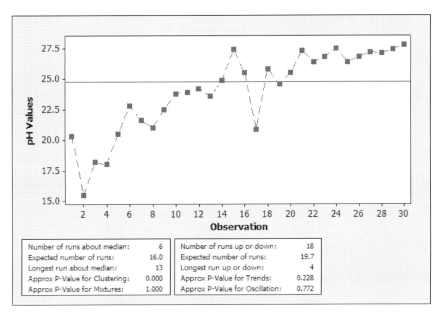

FIGURE 5-5. Run chart of pH values

$$30.4 \pm \frac{(1.96)\,(1.25)\,(32.65 - 28.5)}{1.35\sqrt{50}}, \text{ or}$$

$$30.4 \pm 1.065 = (29.335, 31.465).$$

5-10.   a)   A run chart is constructed using Minitab. The following p-values are indicated: p-value for clustering = 0.000; p-value for mixture = 1.000; p-value for trends = 0.228; p-value for oscillation = 0.772. Hence, since p-value for clustering $< \alpha = 0.05$, there is significant clustering and so the sequence is nonrandom. Figure 5-5 shows the run chart.

   b)   The clustering of observations, by batch, is observed. For examples, for batch 3 (observations 21-30), the pH values are clustered around 26.4 and 27.8.

   c)   As shown in part a), clustering is significant.

   d)   Here stratified random sampling scheme is used where the strata represent the batches. Since mixing of the ingredients takes place by batches, this procedure is appropriate.

5-11.   a)   A run chart is constructed using Minitab. The following p-values are indicated: p-value for clustering = 0.032; p-value for mixtures = 0.968; p-value for trends = 0.117; p-value for oscillation = 0.883. Since p-value for clustering $< \alpha = 0.05$, we conclude that clustering is significant and so the sequence is not random. Figure 5-6 shows the run chart.

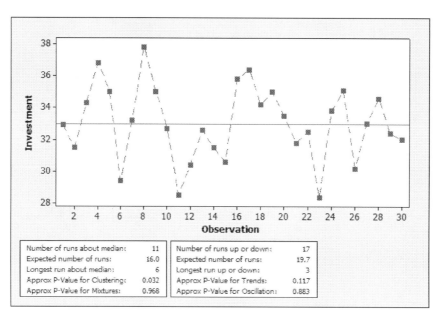

Number of runs about median:	11	Number of runs up or down:	17
Expected number of runs:	16.0	Expected number of runs:	19.7
Longest run about median:	6	Longest run up or down:	3
Approx P-Value for Clustering:	0.032	Approx P-Value for Trends:	0.117
Approx P-Value for Mixtures:	0.968	Approx P-Value for Oscillation:	0.883

FIGURE 5-6. Run chart of percentage investment

b)   p-value for trend = $0.1168 > \alpha = 0.05$; p-value for clustering $< \alpha$. So clustering effect is significant.

c)   Conducting the Anderson-Darling normality test using Minitab, p-value = $0.970 > \alpha = 0.01$. So we do not reject the null hypothesis of normality.

d) $\overline{X} = 33.030$, s = 2.367. A 98% confidence interval for the mean percentage investment is obtained as follows:

$$33.030 \pm t_{0.01,29} \, (2.367 / \sqrt{30})$$

$$= 33.030 \pm 2.462 \, (2.367) / \sqrt{30}$$

$$= 33.030 \pm 1.064 = (31.966, 34.094).$$

e)   A 98% confidence interval for the variance is given by:

$$\frac{29(2.367)^2}{49.6} < \sigma^2 < \frac{29(2.367)^2}{14.3}$$

$$3.276 < \sigma^2 < 11.362.$$

A 98% confidence interval for σ is given by:

$$1.810 < \sigma < 3.371.$$

78

f) $H_o : \mu \leq 31$ vs. $H_a : \mu > 31$. The test statistic is obtained as:

$$t_o = \frac{33.030 - 31}{2.367 / \sqrt{30}} = 4.697$$

The rejection region of H_o is $t_o > t_{0.01,29} = 2.462$. Since the test statistic > 2.462, we reject H_o and conclude that the mean percentage investment exceeds 31%. The p-value $= P[t \geq 4.697] = 0.000$.

5-12. Using last year's data, we calculate the proportion of employees that preferred each plan. These will serve as the test proportions for the current year. The hypotheses to be tested are:

$H_o : p_1 = 0.25, \; p_2 = 0.20, \; p_3 = 0.55$

H_a : At least one p_i differs from the hypothesized value.

$$\text{Test statistic} = X^2 = \frac{(15 - 37.5)^2}{37.5} + \frac{(45 - 30.0)^2}{30} + \frac{(90 - 82.5)^2}{82.5}$$

$$= 21.682$$

Using $\alpha = 0.05$, critical value of $\chi^2_{0.05,2} = 5.99$.

Since the test statistic > 5.99, we reject H_o and conclude that preferences for health care plans have changed from last year.

p-value $= P[\chi^2 > 21.682] = 0.000$.

5-13. H_o : Sales is independent of advertising technique.
   H_a : Sales is not independent of advertising technique.

$$\text{Test statistic} = X^2 = \frac{(60 - 58.33)^2}{58.33} + \frac{(40 - 58.33)^2}{58.33} + \frac{(75 - 58.33)^2}{58.33}$$

$$+ \frac{(85 - 61.67)^2}{61.67} + \dots\dots\dots + \frac{(25 - 40)^2}{40} = 88.038.$$

Using the chi-squared distribution with 6 degrees of freedom, the p-value $= P[\chi^2 \geq 88.038] = 0.000$. Since the p-value $< \alpha = 0.10$, we reject H_o and conclude that the advertising technique has an impact on sales.

5-14. H_o: Overall satisfaction rating is independent of response speed rating.

H_a: Overall satisfaction rating is not independent of response speed rating.

Using the original categories of both variables, it is found that for the category "Response speed rating = 1", two cells (Speed rating = 1, Satisfaction rating = 1) and (Speed rating = 1, Satisfaction rating = 2) have expected frequencies less than 5. Also, for cell (Speed rating = 2, Satisfaction rating = 1), the expected frequency is 4.30 and so is less than 5. To satisfy the assumption of each cell having an expected frequency of at least 5, we collapse rows 1 and 2 (corresponding to speed rating = 1 and 2) and end up with four rows and five columns.

$$\text{Test statistic} = X^2 = \frac{(41-8)^2}{8} + \frac{(33-10.42)^2}{10.42} + \frac{(8-19.53)^2}{19.53}$$

$$+ \frac{(6-21.58)^2}{21.58} + \frac{(5-33.48)^2}{33.48} + \frac{(2-9.98)^2}{9.98} + \cdots$$

$$+ \frac{(70-52.92)^2}{52.92} = 328.396.$$

Using chi-squared distribution with 12 degrees of freedom, the p-value = 0.000. Since the p-value $< \alpha = 0.01$, we reject H_o and conclude that response speed influences overall satisfaction: Cramer's index of association:

$$V = \sqrt{\frac{328.396}{500(4-1)}} = 0.468.$$

5-15.   a)   Sample size is given as:

$$n = z_{0.05}^2 \, (5500)^2/1000^2 = 1.645^2 \, (5500)^2/1000^2 = 81.86 \approx 82.$$

b)   Given $\alpha = .10$, $\sigma = 5500$, difference to detect $(\delta) = 500$ with a power = 0.95, using Minitab, we get $n = 1311$.

5-16.   a)   Since no prior information is available on the proportion of copper in the mixture, to be conservative, we use p = 0.05.

$$n = z_{0.05}^2 \, (0.5) \, (0.5)/(0.04)^2 = 1.645^2 \, (0.25)/(0.04)^2$$

$$= 422.8 \approx 423.$$

b) Given $\alpha = 0.02$, hypothesized value of $p = 0.15$, alternative value of $p = 0.17$, power = 0.98, and the form of the alternative hypothesis being "greater than", using Minitab, we get n = 5661.

5-17. a) $n = 1.645^2 \, (200^2 + 250^2) / 40^2 = 173.35 \simeq 174$.

b) We have the following equation:

$$40 = 1.645 \sqrt{\frac{200^2}{2n_2} + \frac{250^2}{n_2}} = \sqrt{\frac{200^2 + 2(250^{2)}}{2n_2}}$$

or $n_2 = 139.53 \simeq 140$. So $n_1 = 280$.

c) Using an estimate of the standard deviation as the average of the two estimates, yielding a value of 225 minutes, difference to detect = 30 with a power of 0.80, $\alpha = 0.10$, and the form of the alternative hypothesis being "not equal to", using Minitab yields a sample size of 697.

5-18. a) $n = 1.645^2 \, [(0.5)(0.5) + (0.5)(0.5)] / (0.04)^2$

$= 845.6 \simeq 846 = n_1 = n_2$.

b) An estimate of the proportion nonconforming based on observations from the process, before and after changes, when used in the equation for determining sample size, will lead to a smaller sample size. The conservative estimate of 0.5, used in part a), yields the maximum sample size.

c) Using Minitab, we input proportion 1 as 0.08 and proportion 2 as 0.03, a power of 0.8, $\alpha = 0.10$, and the form of the alternative as "greater than". This yields a sample size of 187.

5-19. a) $n = (2.33)^2 \, (0.04)(0.96) / (0.03)^2 = 231.63 \simeq 232$.

b) $n = (2.33)^2 \, (0.5)(0.5) / (0.03)^2 = 1508.03 \simeq 1509$.

5-20. $n = (1.96)^2 \, (5.2)^2 / 2^2 = 25.97 \simeq 26$.

5-21. a) $n_1 = 200(2000) / 10,000 = 40$; $n_2 = 200(7000) / 10,000 = 140$;

$n_3 = 200(1000) / 10,000 = 20$.

b) $$\bar{x}_{st} = \frac{2000(3.5) + 7000(7.6) + 1000(15.1)}{10,000}$$

$= 7.53$ (in \$1000).

Var $(\bar{x}_{st})$ =

$$\frac{1}{10,000^2}\left[2000^2\left(\frac{1960}{2000}\right)\frac{1.2^2}{40} + 7000^2\left(\frac{6860}{7000}\right)\frac{2.8^2}{140} + 1000^2\left(\frac{980}{1000}\right)\frac{6.4^2}{20}\right]$$

$= 0.0484.$

Standard error $(\bar{x}_{st}) = 0.22.$

c) $7.53 \pm (1.96)(0.22) = 7.53 \pm 0.431$ (in \$1000).

d) $6.8 \pm (1.96)(5.6)/\sqrt{200}$ (9800/10,000)

or 6.8 ± 0.7606 (in \$1000).

e) Improvement in precision is found as:

% improvement in precision

$$= \frac{0.396 - 0.220}{0.396} = 44.44\%.$$

5-22. a) Here M = 3, m = 2, $\bar{n}$ = (7000 + 1000) /2 = 4000. Also, t_2 = 7,100,000 and t_3 = 15,200,00.

$$\bar{x}_{cl} = \frac{7,100,000 + 15,200,000}{7000 + 1000} = 2787.5$$

Var $(\bar{x}_{cl}) = \frac{(3-2)}{(3)(2)\,4000^2\,(2-1)} \times$

$$\left[(7,100,100 - 2787.5(7000))^2 + (15,200,000 - 2787.5(1000))^2\right]$$

$= 3209794.922.$

Standard error $(\bar{x}_{cl}) = 1791.59$

b) Bound on the error of estimation

$$= 2787.5 \pm (1.96)(1791.59)$$

$$= 2787.5 \pm 3511.52.$$

c)   There is no representation from the first group. Also, the standard error of the mean for the cluster sample exceeds that of the stratified sample mean.

CHAPTER 6

STATISTICAL PROCESS CONTROL USING CONTROL CHARTS

6-1.   Benefits include when to take corrective action, type of remedial action necessary, when to leave a process alone, information on process capability, and providing a benchmark for quality improvement.

6-2.   Special causes are not inherent in the process. Examples are inexperienced operator, or poor quality raw material and components. Common causes are part of the system. They cannot be totally eliminated. Examples are variations in processing time between qualified operators, or variation in quality within a batch received from a qualified supplier.

6-3.   A normal distribution of the quality characteristic being monitored (for example average strength of a cord) is assumed. For a normal distribution, control limits placed at 3 standard deviations from the mean ensure that about 99.73% of the values will plot within the limits, when no changes have taken place in the process. This implies that very few false alarms will occur.

6-4.   A type I error occurs when we infer that a process is out of control when it is really in control. A type II error occurs when we infer that a process is in control when it is really out of control. The placement of the control limits influences these two errors. As the control limits are placed further out from the center line, the probability of a type I error decreases, but the probability of a type II error increases, when all other conditions remain the same, and vice versa. An increase in the sample size may lead to reducing both errors.

6-5.   Warning limits are these that are placed at 2 standard deviations from the centerline. Using the property of a normal distribution, the probability of an observation falling within the warning/control limit on a given side is about 2.15%, if the process is in control. These limits serve as an alert to the user that the process may be going out of control. In fact, one rule states that if 2 out of 3 successive sample statistics fall within the warning/control limit on a given side, the process may be out of control.

6-6.   The operating characteristic (OC) curve associated with a control chart indicates the ability of the control chart to detect changes in process parameters. It is a measure that indicates the goodness of the chart through its ability to detect changes in the process parameters when there are changes. A typical OC curve for a control chart for the mean will be a graph of the probability of non-detection on the vertical axis versus the process mean on the horizontal axis. As the process mean deviates more from the hypothesized (or current) value, the probability of non-detection should decrease. The discriminatory power of the OC curve may be improved by increasing the sample size.

6-7.   The average run length (ARL) is a measure of goodness of the control chart and represents the number of samples, on average, required to detect an out-of-control signal. For a process in control, the ARL should be high, thus minimizing the number of false alarms. For a process out-of-control, the ARL should be small indicating the sensitivity of the chart. As the degree of shift from the in-control process parameter value increases, the ARL should decrease. Desirable values of the ARL, for both in-control and out-of-control situations, may be used to determine the location of the control limits. Alternatively, from predetermined ARL graphs, the sample size necessary to achieve a desired ARL, for a certain degree of shift in the process parameter, may be determined.

6-8. The ARL is linked to the probability of detection of an out-of-control signal. If P_d represents the probability of detection, we have $ARL = 1/P_d$. For an in-control process, $P_d = \alpha = P$ (type I error). So, for 3σ control limits, $ARL = 1/0.0026 \simeq 385$. For an out-of-control process, $P_d = 1 - P$ (type II error) $= 1 - \beta$. Hence, $ARL = 1/(1-\beta)$.

6-9. The selection of rational samples or subgroups hinges on the concept that samples should be so selected such that the variation within a sample is due to only common causes, representing the inherent variation in the process. Further, samples should be selected such that the variation between samples is able to capture special causes that prevail. Utilization of this concept of rational samples is important in the total quality systems approach since the basic premise of setting up the control limits is based on the inherent variation that exists in the process. Hence, the variability within samples is used to estimate the inherent variation that subsequently impacts the control limits.

6-10. Rule 1 – A single point plots outside the control limits. Rule 2 – Two out of 3 consecutive points plot outside the two-sigma limits on the same side of the centerline. Rule 3 – Four out of 5 consecutive points fall beyond the one-sigma limit on the same side of the centerline. Rule 4 – Nine or more consecutive points fall on one side of the centerline. Rule 5 – A run of 6 or more consecutive points steadily increasing or decreasing. All of the rules are formulated on the concept that, if a process is in control, the chances of the particular event happening is quite small. This is to provide protection against false alarms.

6-11. Some reasons could be adding a new machine, or a new operator, or a different vendor supplying raw material.

6-12. Typical causes could be tool wear in a machining operation or learning on the job associated with an increase in time spent on job.

6-13. Assume that three-sigma control limits are used. For Rule 1, the probability of a type I errors is $\alpha_1 = 0.0026$. For Rule 2, the probability of 2 out of 3 consecutive points falling outside the two-sigma limits, on a given side of the centerline is:

$$\binom{3}{2}(0.0228)^2(0.9772) = 0.001524.$$

Since this can happen on either side, the probability of a type I error using Rule 2 is $\alpha_2 = 2(0.001524) = 0.003048$.

For Rule 3, the probability of 4 out of 5 consecutive points falling beyond the one-sigma limit, on a given side to the centerline, is:

$$\binom{5}{4}(0.1587)^4(0.8413) = 0.002668.$$

Since this can happen on either side, the probability of a type I error using Rule 3 is $\alpha_3 = 2(0.002668) = 0.005336$.

Assuming independence of the rules, the probability of an overall type I error is $\alpha = 1 - (1 - \alpha_1)(1 - \alpha_2)(1 - \alpha_3)$

$$= 1 - (0.9974)(0.996952)(0.994664) = 0.010946.$$

6-14. The centerline on a chart for the average is CL = 15 mm. The standard deviation of the sample mean is $\sigma_{\bar{x}} = \sigma / \sqrt{n} = 0.8 / \sqrt{4} = 0.4$ mm.

a) The one-sigma control limits are: $15 \pm 0.4 = (14.6, 15.4)$.

The two-sigma control limits are: $15 \pm 2(0.4) = (14.2, 15.8)$.

b) The three-sigma control limits are:

UCL $= 15 + 3(0.4) = 16.2$

LCL $= 15 - 3(0.4) = 13.8$.

c) The probability of a false alarm is the probability of a type I error. Using three-sigma limits, the probability of a type I error is 0.0026.

d) The process mean shifts to 14.5 mm. The standardized normal values at the control limits are:

$$Z_1 = \frac{16.2 - 14.5}{0.4} = 4.25 ; \quad Z_2 = \frac{13.8 - 14.5}{0.4} = -1.75.$$

Using the standard normal tables, the area above the UCL is 0.0000, and that below the LCL is 0.0401. The area between the control limits is $1 - (0.0000 + 0.0401) = 0.9599$. Hence the probability of not detecting the shift on the first subgroup plotted after the shift is 0.9599.

ARL $= 1/(1$ - Probability of a type II error)
$= 1/0.0401 = 24.938$.

TABLE 6-1. Computation of Probabilities for OC Curve

Process mean	Z-value at UCL Z_1	Area above UCL	Z-value at LCL Z_2	Area below LCL	Probability of non-detection
15.4	2.00	0.0228	-4.00	0.0000	0.9772
15.8	1.00	0.1587	-5.00	0.0000	0.8413
16.2	0.00	0.5000	-6.00	0.0000	0.5000
16.6	-1.00	0.8413	-7.00	0.0000	0.1587
17.0	-2.00	0.9772	-8.00	0.0000	0.0228
17.4	-3.00	0.9987	-9.00	0.0000	0.0013

e) We first compute the probability of detecting the shift on the first subgroup, which is $(1 - 0.9599) = 0.0401$. Next, the probability of not detecting the shift on first subgroup and detecting shift on second subgroup is $(1 - 0.0401)(0.0401) = 0.0385$, assuming independence of the two subgroups. So, the probability of detecting the shift by the second subgroup is $(0.0401 + 0.0385) = 0.0786$. Hence, the probability of failing to detect the shift by the second subgroup is $(1 - 0.0786) = 0.9214$.

f) Some sample calculations for the coordinates of the OC curve are shown in Table 6-1, assuming that the control limits are at 13.8 and 16.2. Calculations for shifts in the process mean on one side are shown. Similar calculations will hold when the process mean decreases.

g) For values of the process mean in part f), the ARL values are shown in Table 6-2.

6-15. a) The center line on a chart for the average length is 110 mm. The standard deviation of the sample mean is $\sigma_{\bar{x}} = \sigma / \sqrt{n} = 4 / \sqrt{5} = 1.7888$ mm. The warning limits are:

$$110 \pm 2(1.7888) = 110 \pm 3.5776 = (106.4224, 113.5776).$$

b) The three-sigma control limits are:

$$110 \pm 3(1.7888) = 110 \pm 5.3664 = (104.6336, 115.3664).$$

TABLE 6-2. ARL Values for Shifts in Process Mean

Process mean	Probability of non-detection	P_d	ARL
15.4	0.9772	0.0228	43.86
15.8	0.8413	0.1587	6.30
16.2	0.5000	0.5000	2.00
16.6	0.1587	0.8413	1.19
17.0	0.0228	0.9772	1.02
17.4	0.0013	0.9987	1.00

The probability of a type I error is 0.0026.

c)   The process mean shifts to 112 mm. The standardized normal values at the control limits are:

$$Z_1 = \frac{115.3664 - 112}{1.7888} = 1.8819 \approx 1.88$$

$$Z_2 = \frac{104.6336 - 112}{1.7888} = 4.118 \approx -4.12.$$

Using the standard normal tables, the area above the UCL is 0.0301, and that below the LCL is 0.0000. The area between the control limits is 0.9699, which is the probability of non-detection of the shift on a given subgroup.

Probability of detecting shift on first subgroup drawn after the shift = 0.0301. The probability of not detecting the shift on the first subgroup and detecting shift on the second subgroup, assuming independence, is (0.9699)(0.0301) = 0.0292. Similarly, the probability of not detecting the shift on the first and second subgroup and detecting on the third subgroup is (0.9699)(0.9699)(0.0301) = 0.0283. Hence, the probability of detecting the shift by the third sample drawn after the shift is (0.0301 + 0.0292 + 0.0283) = 0.0876.

d)   The chance of detecting the shift for the first time on the second subgroup point drawn after the shift is (0.9699)(0.0301) = 0.0292.

e)   For a shift in the process mean to 112 mm, the probability of detecting the shift on first subgroup drawn after the shift is 0.0301. So ARL = 1/0.0301 = 33.22.

When the process mean changes to 116 mm, the standardized normal values at the control limits are:

$$Z_1 = \frac{115.3664 - 116}{1.7888} = -0.354 \approx -0.35$$

$$Z_2 = \frac{104.6336 - 116}{1.7888} = -6.354 \approx -6.35.$$

Using the standard normal tables, the area above the UCL is 0.6368 and that below the LCL is 0.000. So, the probability of detecting the shift on the first sample is 0.6368. The ARL is obtained as ARL = 1/0.6368 = 1.57, which is, on average, the number of samples required to detect the shift.

90

6-16. a) The center line on a chart for the average tensile strength is 3000 kg. The standard deviation of the sample mean is $\sigma_{\bar{x}} = \sigma / \sqrt{n} = 50 / \sqrt{5} = 22.3607$. The one-sigma control limits are: $3000 \pm 22.3607 = (2977.639, 3022.361)$.

Two sigma control limits are:

$$3000 \pm 2(22.3607) = 3000 \pm 44.7214 = (2955.279, 3044.721).$$

For one-sigma limits, we need to find the probability of a type I error. The probability of an observation plotting outside these limits, if the process is in control, is $2(0.1587) = 0.3174$.

b) Three-sigma control limits are:

$$3000 \pm 3(22.3607) = 3000 \pm 67.0821 = (2932.918, 3067.082).$$

c) Assuming three-sigma control limits, for Rule 1, the probability of a type I error is $\alpha_1 = 0.0026$. For Rule 2, the probability of a type I error (as found in Problem 6-13) is $\alpha_2 = 0.003048$. Assuming independence of the rules, the probability of an overall type I error is:

$$\alpha = 1 - (1 - \alpha_1)(1 - \alpha_2)$$

$$= 1 - (0.9974)(0.996952) = 0.00564.$$

6-17. a) The center line on a chart for the average temperature is 5000°C. The standard deviation of the sample mean is $\sigma_{\bar{x}} = \sigma / \sqrt{n} = 50 / \sqrt{4} = 25$. Three-sigma control limits are:

$5000 \pm 3(25) = (4925, 5075)$.

b) Assuming three-sigma control limits, the probability of a type I error using Rule 2 was previously found (in Problem 6-13) to be 0.003048. Also, for Rule 3, the probability of a type I error was found to be 0.005336. Assuming independence of the rules, the probability of an overall type I error is:

$$\alpha = 1 - (1 - 0.003048)(1 - 0.005336) = 0.008368.$$

c) Since the overall probability of a type I error using Rules 2 and 3 is 0.008368, on average, the number of samples analyzed before an out-of-control condition is indicated is:

$1/0.008368 = 119.5 \simeq 120$.

d)   If the process averages drops to 4960, the standardized normal values at the control limits are:

$$Z_1 = \frac{5075 - 4960}{25} = 4.60 \; ; \; Z_2 = \frac{4925 - 4960}{25} = -1.40.$$

Using the standard normal tables, the area above the UCL is 0.0000, and that below the LCL is 0.0808. The area between the control limits is 0.9192, which is the probability of non-detection of the shift on a given subgroup.

Now, the probability of detecting shift on the first subgroup is 0.0808. Next, the probability of not detecting the shift on the first subgroup and detecting on the second subgroup is $(0.9192)(0.0808) = 0.0743$. Similarly, the probability of not detecting the shift on the first two subgroups and detecting on the third subgroup is $(0.9192)(0.9192)(0.0808) = 0.0683$, assuming independence of the subgroups. Hence, the probability of detecting the change by the third subgroup is $(0.0808 + 0.0734 + 0.0683) = 0.2234$. Thus, the probability of failing to detect the change by the third subgroup point drawn after the change is $(1 - 0.2234) = 0.7766$.

e)   Using the computations in part d) of this problem, the probability of the shift being detected within the first two subgroups is $(0.0808 + 0.0743) = 0.1551$.

6-18. a)   For Rule 1, the probability of a type I error is $\alpha_1 = 0.0026$. For Rule 4, the probability of 8 consecutive points falling on one side of the center line is $(0.5)^8 = 0.003906$. Since this can happen on either side of the center line, the probability of a type I error is $\alpha_4 = 2(0.003906) = 0.007812$. Therefore, the probability of an overall type I error is:

$$\alpha = 1 - (1 - 0.0026)(1 - 0.007812) = 0.01039.$$

b)   When the process mean was at 105 mm, the three-sigma control limits are calculated as:

$$105 \pm 3(6/\sqrt{4}) = 105 \pm 9 = (96, 114).$$

With the process mean at 110 mm, using Rule 1, let us calculate the probability of a subgroup mean plotting outside the control limits. The standardized normal values at the control limits are:

$$Z_1 = \frac{114 - 110}{3} = 1.333 \simeq 1.33; \; Z_2 = \frac{96 - 110}{3} = -4.667 \simeq -4.67.$$

Using the standard normal tables, the area above the UCL is 0.0918, and that below the LCL is 0.0000. Thus the probability of an observation plotting outside the control limits is 0.0918.

With the process mean at 110 mm, the probability of a subgroup mean plotting above the center line is calculated as follows. The standardized normal value at the center line is:

$$Z = \frac{105 - 110}{3} = -1.666 \simeq -1.67.$$

The area above the center line of 105 is $(1 - 0.0475) = 0.9525$. Now, using Rule 4, the probability of 8 consecutive observations plotting above the centerline, assuming independence, is $(0.9525)^8 = 0.677516$. The probability of an observation plotting below the centerline is $(1 - 0.9525) = 0.0475$. As before, the probability of 8 consecutive observations plotting below the centerline is $(0.0475)^8 = 2.59 \times 10^{-11}$, which is negligible. Hence the probability of 8 consecutive observations falling on one side of the centerline is 0.677516. Assuming independence of the two rules, the probability of an out-of-control condition is:

$$1 - (1 - 0.0918)(1 - 0.677516) = 0.70712.$$

Therefore, on average, the number of subgroups analyzed before detecting a change is $1/0.70712 = 1.414 \approx 2$. But since Rule 4 can indicate an out-of-control condition with a minimum of 8 observations, the average number of subgroups needed would be 8. Note that if only Rule 1 were used, on average, $1/0.0918 = 10.89 \approx 11$ subgroups would be needed before detecting a change.

6-19. a) The center line on a chart for the average delivery time is 140 hours. The standard deviation of the sample mean is $\sigma_{\bar{x}} = \sigma / \sqrt{n} = 6 / \sqrt{4} = 3$. The two-sigma control limits are:

140 ± 2(3) = (134, 146).

The three-sigma limits are:

140 ± 3(3) = (131, 149).

b) A type I error, in this context, implies concluding that the average delivery time differs from 140 hours, when in fact it is equal to 140 hours. A type II error implies concluding that the average delivery time is 140 hours, when in fact it differs from 140 hours.

c)   From the results in Problem 6-13, the probability of a type I error using Rule 1 is $\alpha_1 = 0.0026$ and the probability of a type I error using Rule 3 is $\alpha_3 = 0.005336$. Assuming independence of the rules, the probability of an overall type I error is:

$$\alpha = 1 - (1 - \alpha_1)(1 - \alpha_3)$$

$$= 1 - (1 - 0.0026)(1 - 0.005336) = 0.007922.$$

d)   The mean delivery time shifts to 145 hours. Using only Rule 1, we demonstrate calculation of the probability of detection:

$$Z_1 = \frac{149 - 145}{3} = 1.33; \ Z_2 = \frac{131 - 145}{3} = -4.67.$$

Using the standard normal tables, the area above the UCL is 0.0918, while the area below the LCL is 0.0000. The area between the control limits is 0.9082, which is the probability of non-detection of the shift on a given subgroup.

Now, the probability of detecting shift on the first subgroup is 0.0918. The probability of not detecting the shift on the first subgroup and detecting on the second subgroup is $(0.9082)(0.0918) = 0.083373$. Hence, the probability of detecting the shift by the second sample is $(0.0918 + 0.083373) = 0.1752$. Thus, the probability of not detecting the shift by the second sample is $(1 - 0.1752) = 0.8248$.

e)   ARL $= 1/0.0918 = 10.89$. On average, if using only Rule 1, if the process mean shifts to 145, it will take about 10.89 samples to detect this change.

6-20. a)   Note that individual expenditures are being monitored. The three-sigma control limits are:

$15 \pm 3(2) = (9, 21)$ in \$100.

b)   From the results in Problem 6-13, the probability of a type I error using Rule 1 is $\alpha_1 = 0.0026$ and the probability of a type I error using Rule 2 is $\alpha_2 = 0.003048$. Assuming independence of the rules, the probability of an overall type I error is:

$$\alpha = 1 - (1 - \alpha_1)(1 - \alpha_2)$$
$$= 1 - (1 - 0.0026)(1 - 0.003048) = 0.00564.$$

A type I error in this context implies concluding that the mean expenditure differs from \$1500, when in fact it does not.

c)   We demonstrate calculation of the probability of detection, when only rule 1 is used:

$$Z_1 = \frac{21 - 17.50}{2} = 1.75; \quad Z_2 = \frac{9 - 17.50}{2} = -4.25.$$

Using the standard normal distribution, the area above the UCL is 0.0401, while the area below the LCL is 0.0000. The area between the control limits is 0.9599, which is the probability of non-detection of the shift on a given sample.

The probability of detecting the shift by the second sample
$= 0.0401 + (0.9599)(0.0401) = 0.0401 + 0.0385 = 0.0786.$

d)   ARL = 1/0.0401 = 24.94.

CHAPTER 7

CONTROL CHARTS FOR VARIABLES

7-1.	Variables provide more information then attributes since attributes do not show the degree of conformance. Variables charts are usually applied at the lowest level (for example operator or machine level). Sample sizes are typically smaller for variables charts. The pattern of the plot may suggest the type of remedial action, if necessary, to take. The cost of obtaining variable data is usually higher than that for attributes.

7-2.	The Pareto concept is used to select the "vital few" from the "trivial many" characteristics that may be candidates for selection of monitoring through control charts. The Pareto analysis could be done based on the impact to company revenue. Those characteristics that have a high impact on revenue could be selected.

7-3.	A variety of preliminary decisions are necessary. These involve selection of rational samples, sample size, frequency of sampling, choice of measuring instruments, and design of data recording forms as well as the type of computer software to use. In selecting rational samples, effort must be made to minimize variation within samples such that it represents the inherent variation due to common causes that exists in the system. Conversely, samples must be so chosen to maximize the chances of detecting differences between samples, which are likely due to special causes.

7-4.	a)	Data to collect would be the waiting time of passengers to check in baggage. Depending on the sample size, an appropriate control chart will be constructed as follows: For small sample sizes (≤ 10) an $\bar{X}$ and R chart could be used, while for large sample sizes (>10) an $\bar{X}$ and s chart may be appropriate. If data on individual passengers is collected (n = 1), rather than in subgroups, charts for individuals (I) and moving range (MR) could be monitored.

b)	For n $\leq$ 10, use an $\bar{X}$ and R chart, where product assembly time data is chosen randomly. For n >10, an $\bar{X}$ and s chart may be used. For data on individual assembly time (n = 1), an I and MR chart could be monitored.

c)	For individuals data, an I and MR chart could be used.

d)	For n $\leq$ 10, and $\bar{X}$ and R chart could be used. Data on emission levels, say in ppm, could be collected.

e)	A cumulative sum chart, a moving average chart, or an exponentially weighted moving average chart may be used.

f)	For individuals data, an I and MR chart could be used.

g)	A trend chart (or regression control chart) may be used.

h)	An acceptance control chart may be used.

7-5.	One has to be careful in drawing conclusions from a control chart based on standard values. The process could indicate "signs of out-of-control" conditions, through plotting

98

of observations outside the control limits, for example, when there may not be special causes. It could be that the process is in control but not capable of meeting the imposed standards. In this situation, management will need to address the common causes and identify means of process improvement.

7-6.   If the process standard deviation for each product is approximately the same, an $\bar{X}$ and R chart for short production runs, where the deviation of the observed value from a specified nominal value, for each characteristic, is monitored.   Alternatively, a standardized control chart (Z and MR) chart could be used. For each characteristic, a standardized value ((Observed value – mean)/standard deviation) is obtained for each observation.

7-7.   Since there are multiple quality characteristics, not independent of each other, with target values for each specified, a multivariate control chart, such as a Hotelling's T^2 chart, with individual observations could be used.

7-8.   On an $\bar{X}$ chart, if an observation falls below the LCL, it implies an unusually fast response to a fire alarm. For an R chart, an observation plotting below the LCL implies that the spread in the response time is small. For the $\bar{X}$ chart, the point plotting below the LCL is rather desirable.   Hence, if we can identify the special conditions that facilitated its occurrence, we should attempt to adopt them. If such is feasible, we may not delete that observation during the revision process. For the observation below the LCL on the R-chart, it implies the process variability to be small for that situation. Reducing variation is a goal for everyone. Thus, it may be worthwhile looking into conditions that led to its occurrence and emulating them in the future. If this is feasible, we may not delete the observation during the revision process.

7-9.   On the $\bar{X}$-chart, we might expect some oscillations around the centerline, with the magnitude of the oscillations reducing with time on the job. On the R-chart, we would expect a downward trend showing a gradual decrease in the range values as learning on the job takes place.

7-10.   To reduce average preparation time, one could identify the distinct components of the proposal. Personnel could then be assigned to each component and trained accordingly to complete that segment in an efficient manner. Factors that impede flow from one unit to the next, as the proposal is completed, could be investigated and actions taken to minimize bottlenecks. To reduce the variability of preparation times, tasks could be standardized to the extent possible. Further, to reduce the average and the variability, use of a common database, could be explored. Thus, one component of a proposal need not wait for the completion of another part, unless the final results from the first part are absolutely necessary.

7-11.   When using an individual chart, the major assumption when using 3-sigma limits is normality of distribution of the characteristic. In this case it is the patient recovery time. If the distribution is non-normal, the particular distribution that fits the observations should be used to determine the centerline and control limits.

7-12. The capability of a process (its ability to meet specifications or customer requirements) should be estimated only after it has been brought to the state of statistical control. Thus, at this point only common causes prevail in the system, indicating the inherent variability in the process. In this context, it could mean the ability to complete the construction of the office building within a defined period. With special causes having been addressed, we are in a better position to forecast timely completion of the construction project.

7-13. A major advantage of cumulative sum charts compared to Shewhart control charts is the ability to detect small shifts in the process mean. Disadvantages could be the complexity of the chart, and it could be slow to detect large shifts. Also, cumulative sum charts may not be effective to study past performance and determine if the process is in control or needs to be brought into control. They are used for stable processes.

7-14. A moving-average control chart is desirable when it is preferred to detect small shifts in the process mean. A geometric moving-average chart is also used to detect small changes in the process mean, and may be more effective than the moving-average chart. The geometric moving-average chart assigns more weight to the more recent observations. The weights decrease exponentially for observations that are less recent.

7-15. In a trend control chart, the quality characteristic being monitored is expected to increase or decrease gradually rather than remain at a constant level. For example, with tool or die wear, machined characteristic may follow this pattern. Also, it is assumed the specification spread is wider than the inherent process spread, which allows for the quality characteristic to drift (upward or downward) but still produce parts that satisfy specifications.

7-16. In a modified control chart, the objective is to determine bounds on the process mean such that the proportion of nonconforming items does not exceed a specified desirable value. Further, the probability of a false alarm is to not exceed a given level. In an acceptance control chart, the objective is to determine the bounds on the process mean, such that we wish to detect a specified level of proportion nonconforming with a desired probability. For both charts, the common assumptions are that the inherent process spread is much less then the specification spread, the process variability is in control, and the distribution of the individual quality characteristic is normal.

7-17. When there are several quality characteristics, that are not necessarily independent of each other, and need to be monitored for control of the process/product, multivariate control charts deserve attention. Errors of both types (type I and type II) could be reduced by using multivariate control charts relative to control charts for each individual characteristic monitored separately. The individual charts may not incorporate the relationship of that characteristic with others. Depending on the number of quality characteristics being monitored jointly, it is known that if individual charts are kept, the overall type I error could be quite high.

7-18. Multivariate control chart (say Hotelling's T^2 chart) where individual observations are chosen on each characteristic (for example, age, respiration rate, heart rate, temperature, pulse oximetry) for individual infants.

7-19. a) The trial control limits for $\bar{X}$ and R charts are shown in Figure 7-1. On the $\bar{X}$-chart, the trial limits are (353.382, 346.472), with the centerline being 349.927. On the R-chart, the trial control limits are (10.82, 0), with the centerline being 4.74.

b) Sample numbers 9 and 11 plot outside the $\bar{X}$-chart control limits. Assuming special causes and appropriate remedial actions, they are deleted while calculating the revised control limits. On the $\bar{X}$-chart, the revised limits are (353.431, 346.432), with the centerline being 349.932. On the R-chart, the revised limits are (10.96, 0), with the centerline being 4.80. All of the points are now within the control limits. Figure 7-2 shows the revised control limits.

c) Specifications are 350 ± 5. An estimate of the process standard deviation is $\hat{\sigma} = \bar{R}/d_2 = 4.80/2.059 = 2.331$. The standard normal values at the specification limits are:

$$Z_1 = (345 - 349.932)/2.331 = -2.12; \quad Z_2 = (355 - 349.932)/2.331 = 2.17.$$

The proportion below LSL is 0.0170, while the proportion above USL is (1 − 0.9850) = 0.0150. The total proportion of nonconforming bottles is 0.032. The number of bottles nonconforming daily is 20,000 (0.032) = 640.

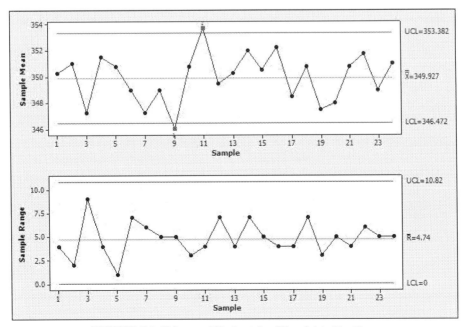

FIGURE 7-1. X-bar and R chart for fill weight of bottles

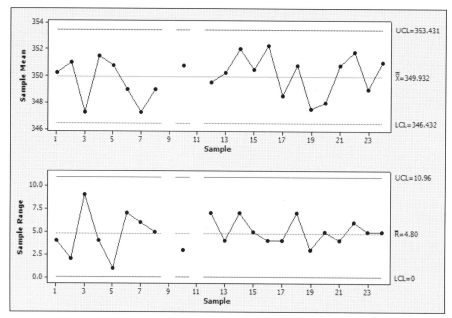

FIGURE 7-2. Revised X-bar and R chart for fill weight of bottles

d) Daily cost of rectifying under-filled bottles = (20,000)(0.0170)(0.08) = $27.20. Daily loss of revenue on over-filled bottles = (20,000)(0.0150)(0.03) = $9.00. Hence, daily revenue loss, on average = $36.20. Assuming 30 days in a month, the monthly revenue lost on average = $1086.

e) The standard normal values at the $\bar{X}$-chart control limits, when the process average is at 342 are:

$$Z_1 = (346.432 - 342)/(2.331/\sqrt{4}) = 3.80;$$

$$Z_2 = (353.431 - 342)/(2.331/\sqrt{4}) = 9.81.$$

The probability of the sample mean plotting below the LCL is approximately 0.9999276520, while the probability of plotting above the UCL is about 0.0000. So, the probability of detecting the shift on the next sample drawn after the shift is 0.999927652.

f) To find the proportion nonconforming when the process average is at 342, the standard normal values at the specification limits are:

$$Z_1 = (345 - 342)/2.331 = 1.29; \quad Z_2 = (355 - 342)/2.331 = 5.58.$$

The proportion nonconforming below the LSL is 0.9015, while that above the USL is 0.0000. Hence, about 90.15% of the product will be nonconforming.

7-20. a) $\overline{\overline{X}} = 1000/25 = 40$, $\overline{R} = 250/25 = 10$. The control limits on an R-chart are: UCL = 2.282(10) = 22.82, LCL = 0(10) = 0. The control limits on an $\overline{X}$ chart are: UCL = 40 + 0.729(10) = 47.29, LCL = 40 − 0.729(10) = 32.71.

  b)   An estimate of the process standard deviation is $\hat{\sigma} = \overline{R}/d_2 = 10/2.059 = 4.8567$. The standard normal value at 50 is Z = (50 − 40)/4.8567 = 2.059 ≃ 2.06. The proportion of customers who will not wait more than 50 minutes is 0.9803.

  c)   The two-sigma control limits for the $\overline{X}$-chart are:

$$40 \pm 2(4.8567/\sqrt{4}) = 40 \pm 4.8567 = (35.143, 44.857).$$

  d)   The process average waiting time is reduced to 30 minutes. Standard normal value at 40 is Z = (40 − 30)/4.8567 = 2.059 ≃ 2.06. Proportion of customers who will have to wait more than 40 minutes is (1 − 0.9803) = 0.0197. Standard normal value at 50 now is Z = (50 − 30)/4.8567 = 4.118, with the proportion of customers who will have to wait more than 50 minutes being negligible (0.0000).

7-21. $\overline{X}$ and R charts are constructed using the given data. We have:

$$\overline{\overline{X}} = \frac{241.4}{25} = 9.656; \quad \overline{R} = \frac{89.8}{25} = 3.592.$$

The trial control limits on the R-chart are:

UCL = 2.114(3.592) = 7.593; LCL = 0(3.592) = 0.

The trial control limits on the $\overline{X}$-chart are:

UCL = 9.656 + 0.577(3.592) = 11.729

LCL = 9.656 − 0.577(3.592) = 7.583.

No values are outside the control limits on the R-chart. On the $\overline{X}$-chart, observations 1 and 6 plot below the lower control limit and observation 3 plots above the upper control limit. Since the first two cases have small delay times, which is desirable, assuming that process circumstances under those situations can be adopted in the future, we delete observation 3 and recalculate the limits.

The revised centerline on the $\overline{X}$-chart is 9.400, and the revised centerline on the R-chart is 3.550. The revised control limits on the R-chart are:

UCL = 2.114(3.550) = 7.505; LCL = 0(3.550) = 0.

The revised control limits on the $\overline{X}$-chart are:

UCL = 9.400 + 0.577(3.550) = 11.448; LCL = 9.400 − 0.577(3.550) = 7.352.

The average delay time is 9.4 minutes. Since the goal is not to exceed 10 minutes, we calculate the probability of meeting this goal. The standard deviation of delay times is estimated as:

$$\hat{\sigma} = \frac{3.550}{2.326} = 1.526.$$

The standard normal value at 10 is:

$$Z = (10 − 9.4)/1.526 = 0.393 \approx 0.39.$$

Using the standard normal tables, the probability of a delay of 10 minutes or less is 0.6517, or about 65%. Hence, the airline must strive to reduce delay times further since it will not meet the goal about 35% of the time.

7-22. a) $\overline{\overline{X}}$ = 195/25 = 7.8, $\overline{R}$ = 10/25 = 0.4. The control limits on an R-chart are: UCL = 2.282(0.4) = 0.9128, LCL = 0(0.4) = 0. The control limits on an $\overline{X}$-chart are: UCL = 7.8 + 0.729(0.4) = 8.0916, LCL = 7.8 − 0.729(0.4) = 7.5084.

b) The one-sigma limits on an $\overline{X}$-chart are:

7.8 ± 0.729(0.4)/3 = 7.8 ± 0.0972 = (7.703, 7.877).

Two-sigma limits on an $\overline{X}$-chart are:

7.8 ± 2(0.0972) = 7.8 ± 0.1944 = (7.606, 7.944).

c) An estimate of the process standard deviation is $\hat{\sigma} = \overline{R}/d_2$ = 0.4/2.059 = 0.194. Specifications are 7.5 ± 0.5 = (7, 8). The standard normal values at the specification limits are as follows:

Z_1 = (7 − 7.8)/0.194 = − 4.12; Z_2 = (8 − 7.8)/0.194 = 1.03.

The fraction of the output below the lower specification limit is negligible (0.0000), while that above the upper specification limits is (1 − 0.8485) = 0.1515.

7-23. a) $\overline{\overline{X}}$ = 107.5/25 = 4.3, $\overline{R}$ = 12.5/25 = 0.5. The control limits on an R-chart are: UCL = 2.282(0.5) = 1.141, LCL = 0(0.5) = 0. The control limits on an $\overline{X}$-chart are: UCL = 4.3 + 0.729(0.5) = 4.664, LCL = 4.3 − 0.729(0.5) = 3.936.

b)   An estimate of the process standard deviation is $\hat{\sigma} = \bar{R}/d_2 = 0.5/2.059 = 0.2428$.

c)   The standard normal values at the specification limits are:

$Z_1 = (4.2 - 4.3)/0.2428 = -0.41$; $Z_2 = (4.6 - 4.3)/0.2428 = 1.2356 \simeq 1.24$.

The proportion below the LSL, representing rework since the dimension is the bore size, is 0.3409. The proportion above the USL, representing scrap, is $(1 - 0.8925) = 0.1075$.

d)   Daily cost of rework = 1200 x 0.3409 x 0.75 = \$306.81.
Daily cost of scrap = 1200 x 0.1075 x 2.40 = \$309.60.

e)   If the process average shifts to 4.5 mm, the standard normal values at the specification limits are recalculated as follows:

$Z_1 = (4.2 - 4.5)/0.2428 = -1.236 \simeq -1.24$

$Z_2 = (4.6 - 4.5)/0.2428 = 0.412 \simeq 0.41$.

The proportion below the LSL, representing rework, is now 0.1075, while the proportion above the USL, representing scrap is $(1 - 0.6591) = 0.3409$.

Daily cost of rework now = 1200 x 0.1075 x 0.75 = \$96.75.
Daily cost of scrap now = 1200 x 0.3409 x 2.40 = \$981.79.

7-24. a) $\bar{\bar{X}} = 306/30 = 10.2$, $\bar{R} = 24/30 = 0.8$. The control limits on an R-chart are: UCL = 2.114(0.8) = 1.691, LCL = 0(0.8) = 0. The control limits on an $\bar{X}$-chart are: UCL = 10.2 + 0.577(0.8) = 10.662, LCL = 10.2 − 0.577(0.8) = 9.738.

b)   The one-sigma limits for the $\bar{X}$-chart are:

10.2 ± 0.577(0.8)/3 = 10.2 ± 0.1539 = (10.046, 10.354).

The two-sigma limits for the $\bar{X}$-chart are:

10.2 ± 2(0.1539) = 10.2 ± 0.3078 = (9.892, 10.508).

c)   An estimate of the process standard deviation is $\hat{\sigma} = \bar{R}/d_2 = 0.8/2.326 = 0.344$. The standard normal value at 10.5 minutes is:

$Z = (10.5 - 10.2)/0.344 = 0.872 \simeq 0.87$.

The proportion above 10.5 is $(1 - 0.8078) = 0.1922$, implying that about 19.22% of the customers will leave.

7-25.   a) $\overline{\overline{X}} = 199.89/20 = 9.9945$, $\overline{s} = 2.87/20 = 0.1435$. The control limits on an s-chart are: $UCL = B_4\overline{s} = 2.266(0.1435) = 0.325$, $LCL = B_3\overline{s} = 0(0.1435) = 0$. For sample number 14, the standard deviation is above the UCL. The revised centerline is $\overline{s} = (2.87 - 0.34)/19 = 0.133$. The revised control limits on the s-chart are: $UCL = 2.266(0.133) = 0.301$, $LCL = 0(0.133) = 0$.

Deleting sample number 14, $\overline{\overline{X}} = (199.89 - 10.18)/19 = 9.985$. The control limits for an $\overline{X}$-chart are: $UCL = 9.985 + 1.628(0.133) = 10.201$, $LCL = 9.985 - 1.628(0.133) = 9.768$. The averages of sample numbers 6 and 16 are below the LCL, while that of sample number 4 is above the UCL. The revised centerline on the chart is $\overline{X} = (189.71 - 10.54 - 9.45 - 9.57)/16 = 10.009$. The revised centerline on the s-chart is $\overline{s} = (2.53 - 0.19 - 0.09 - 0.09)/16 = 0.135$. The revised control limits for the $\overline{X}$-chart are: $UCL = 10.009 + 1.628(0.135) = 10.229$, $LCL = 10.009 - 1.628(0.135) = 9.789$. All of the observations are now in control.

b)   The process mean is estimated from the revised centerline on the $\overline{X}$ chart and is 10.009. An estimate of the process standard deviation is $\hat{\sigma} = \overline{s}/c_4 = 0.135/0.9213 = 0.1465$.

c)   The standardized normal values at the specification limits are:

$Z_1 = (9.65 - 10.009)/0.1465 = -2.45$; $Z_2 = (10.25 - 10.009)/0.1465 = 1.645$.

The proportion of the product below the LSL indicating scrap is 0.0071, while the proportion above the USL indicating rework is 0.0500.

Daily cost of rework = 4 x 100 x 80 x 0.0500 x 0.25 = $400.00.
Daily cost of scrap = 4 x 100 x 80 x 0.0071 x 0.75 = $170.40.

d)   If the process mean is moved to 10.00, the standard normal values at the specification limits are:

$Z_1 = (9.65 - 10.00)/0.1465 = 2.389 \simeq 2.39$

$Z_2 = (10.25 - 10.00)/0.1465 = 1.706 \simeq 1.71$.

The proportion of scrap now is 0.0084, while the proportion of rework is 0.0436.

Daily cost of rework = 4 x 100 x 80 x 0.0436 x 0.25 = $348.80.

Daily cost of scrap = 4 x 100 x 80 x 0.0084 x 0.75 = $201.60.

A decrease in the total cost occurs, and so the process mean should be moved to 10.00.

e)   Cause and effect analysis could reveal changes in process parameter settings that might reduce the process variability of the thickness of sheet metal. Vendor control could also be pursued if the quality of in-coming raw material is not consistent and acceptable.

7-26.   a) $\bar{\bar{X}}$ = 2550/30 = 85, $\bar{s}$ = 195/30 = 6.5. The control limits for an s-chart are: UCL = 2.089(6.5) = 13.578, LCL = 0(6.5) = 0. The control limits for an $\bar{X}$-chart are: UCL = 85 + 1.427(6.5) = 94.2755, LCL = 85 − 1.427(6.5) = 75.7245.

b)   The process mean is estimated as 85. The process standard deviation is estimated as $\hat{\sigma} = \bar{s} / c_4$ = 6.5/0.9400 = 6.915.

c)   The standardized normal values at the specification limits are:

Z_1 = (75 − 85)/6.915 = − 1.446 ≃ − 1.45; Z_2 = (105 − 85)/6.915 = 2.892 ≃ 2.89.

The proportion of the product below the LSL is 0.0735, while the proportion above the USL is 0.0019, making the proportion nonconforming as 0.0754.

d)   If the process mean moves to 90, the standard normal values at the specification limits are:

Z_1 = (70 − 90)/6.915 = − 2.169 ≃ − 2.17; Z_2 = (105 − 90)/6.915 = 2.169 ≃ 2.17.

The proportion of the product below the LSL is 0.0150, while that above the USL is also 0.0150, making the proportion nonconforming as 0.0300. A reduction is achieved in the proportion nonconforming. So, in this case, centering the process mean to 90 improves the performance of the process. Other measures might include process analysis to determine suitable process parameter settings that may reduce process variability.

7-27.   a) $\bar{\bar{X}}$ = 398/25 = 15.92, $\bar{s}$ = 3.00/25 = 0.12. The control limits for an s-chart are UCL = 1.815(0.12) = 0.2718, LCL = 0.185(0.12) = 0.0222. The control limits for an $\bar{X}$-chart are: UCL = 15.92 + 1.099(0.12) = 16.0519, LCL = 15.92 − 1.099(0.12) = 15.7881.

b)   The process mean is estimated as 15.92. The process standard deviation is estimated as $\hat{\sigma}$ = 0.12/0.9650 = 0.1243.

c) The distance between the centerline and the UCL for the s-chart is 0.0978. So, the one-sigma control limits on the s-chart are:

$$0.12 \pm 0.0978/3 = 0.12 \pm 0.0362 = (0.0874, 0.1526).$$

Similarly, the two-sigma control limits on the s-chart are:

$$0.12 \pm 2(0.0978)/3 = 0.12 \pm 0.0652 = (0.0548, 0.1852).$$

For the $\bar{X}$-chart, the distance between the centerline and UCL is 0.1319. So, the one-sigma limits on the $\bar{X}$-chart are:

$$15.92 \pm 0.1319/3 = 15.92 \pm 0.0440 = (15.876, 15,964).$$

The two-sigma limits on the $\bar{X}$-chart are:

$$15.92 \pm 2(0.1319)/3 = 15.92 \pm 0.0880 = (15.832, 16.008).$$

d) Standardized normal values at the specification limits are:

$$Z_1 = (15.7 - 15.92)/0.1243 = -1.77$$

$$Z_2 = (16.3 - 15.92)/0.1243 = 3.06.$$

The proportion of the product below the LSL is 0.0384, while that above the USL is 0.0011, making the proportion nonconforming as 0.0395. The process is therefore not completely capable.

e) The standard normal value at the advertised weight of 16 ounces is: $Z = (16 - 15.92)/0.1243 = 0.64$. The proportion of the product below this advertised weight is 0.7389.

f) One measure could be to increase the process mean from its current setting of 15.92 to three standard deviations above the advertised weight of 16.00, which is $16.00 + 3(1.243) = 16.3729$. This will lead to a negligible proportion of the output (0.0013) to be below the advertised weight. Other measures could involve reducing the process variability.

7-28. a) $\bar{\bar{X}} = 199.8/20 = 9.99$, $\bar{s} = 1.40/20 = 0.07$. The control limits for an s-chart are: UCL = 1.970(0.07) = 0.1379, LCL = 0.030(0.07) = 0.0021. The control limits for an $\bar{X}$-chart are: UCL = 9.99 + 1.287(0.07) = 10.080, LCL = 9.99 − 1.287(0.07) = 9.900.

b)   An estimate of the process mean is 9.99, and an estimate of the process standard deviation is $\hat{\sigma} = 0.07/0.9515 = 0.0736$.

c)   Standardized normal values at the specification limits are:

$Z_1 = (9.8 - 9.99)/0.0736 = -2.58$

$Z_2 = (10.2 - 9.99)/0.0736 = 2.85$.

The proportion of the product below the LSL is 0.0049, while that above the USL is 0.0022, making the proportion nonconforming as 0.0071.

d)   If the process mean shifts to 10, the standardized normal values at the specification limits are:

$Z_1 = (9.8 - 10)/0.0736 = -2.717 \simeq -2.72$

$Z_2 = (10.2 - 10)/0.0736 = 2.717 \simeq 2.72$.

The proportion of the product below the LSL is 0.0033, with the same proportion being above the USL, making the proportion nonconforming as 0.0066. A slight reduction is achieved in the proportion nonconforming.

e)   If the process mean changes to 10.2, the standardized normal values at the control limits for the sample average are:

$Z_1 = (9.90 - 10.2)/(0.0736/\sqrt{6}) = -9.98$

$Z_2 = (10.08 - 10.2)/(0.0736/\sqrt{6}) = -3.99$.

The area below the LCL is negligible (0.0000), while the area above the UCL is very close to 1. Therefore, the chance of a subgroup average plotting outside the control limits is very close to 1, making the probability of detection of the shift in the process mean very close to 1.

7-29.   From the data, $\overline{\overline{X}} = 166.5/25 = 6.66$, $\overline{R} = 45.5/25 = 1.82$. The R-chart control limits are:

UCL = (2.282)(1.82) = 4.153

LCL = 0(1.82) = 0.

All of the observations are within the control limits.

The $\overline{X}$-chart limits are:

UCL = 6.66 + 0.729(1.82) = 7.987

LCL = 6.66 – 0.729(1.82) = 5.333.

The characteristic of the amount of dissolved oxygen is one where larger is better. Observation numbers 2, 10, 17, 18, 20, and 22 plot above the UCL, which are desirable. The observation numbers 13 and 14 plot below the LCL and are undesirable. Assuming special causes for the out-of-control points have been identified, we delete the two undesirable observations and revise the limits. For the R-chart:

Revised centerline = 41/23 = 1.783.

UCL = (2.282)(1.783) = 4.069

LCL = 0(1.783) = 0.

For the $\bar{X}$ -chart:

Revised centerline = 158/23 = 6.870.

UCL = 6.870 + 0.729(1.783) = 8.170

LCL = 6.870 – 0.729(1.783) = 5.570.

An estimate of the process standard deviation is $\hat{\sigma}$ = 1.783/2.059 = 0.866. The process average is estimated to be 6.870. So, the standard normal value at the prescribed standard of 4 parts per million is:

Z = (4 – 6.870)/0.866 = – 3.31.

The probability of exceeding the standard is 1 – 0.0005 = 0.9995, which is quite high. Thus, the standards are being achieved.

7-30. A control chart for individuals has a centerline $\bar{X}$ = 88.56, UCL = 99.48, LCL = 77.64. A chart for the moving range of two consecutive observations has a centerline $\overline{MR}$ = 4.105, UCL = 13.41, LCL = 0. From Figure 7-3, none of the observations are outside the control limits.

7-31. It is given that σ = 0.2, α = 0.05, δ = 0.15 in raw units. Choosing the allowable slack as halfway between the target value and the shifted value, K = 0.15/2 = 0.075.

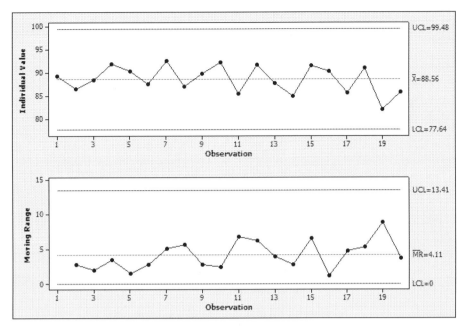

FIGURE 7-3. Individuals and moving range chart for octane rating

We have: $\sigma_{\bar{x}} = \dfrac{0.2}{\sqrt{4}} = 0.1$, which we use as the standard deviation in constructing the cumulative sum chart. We find the slack parameter (k) in standard deviation units = 0.075/0.1 = 0.75, and select the parameter of decision interval (h) in standard deviation units = 5. The cumulative sum chart is shown in Figure 7-4, using the target value as 80.0, for a one-chart. An upward trend in the mean is first detected on Sample 6. Similarly, for detecting downward shifts, this is first detected on Sample 8. A two-sided chart, using a V-mask, if used, also detects a shift on Sample 6.

7-32. L(0) = 300, δ = 0.75, σ = 0.8, μ_o = 30. The parameters of a V-mask are found using Table 7-9.

Lead distance (d) = 15.0, L(0.75) = 14.5,

$\quad$ (k/$\sigma_{\bar{x}}$) tan θ = 0.375.

Using $k = 2\sigma_{\bar{x}}$, we get:

$\quad$ tan θ = 0.375/2 = 0.1875, or θ = 10.62°.

If the manufacturer desires L(0.75) to not exceed 13, we select the following:

$\quad$ Lead distance (d) = 11.3, L(0.75) = 11.0,

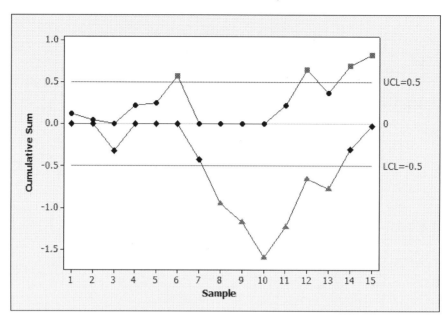

FIGURE 7-4. Cumulative sum chart for average weight

$$(k / \sigma_{\bar{x}}) \tan \theta = 0.375.$$

Using $k = 2\sigma_{\bar{x}}$, yields $\theta = 10.62°$. However, L(0) is now 100.

7-33. The difference between the upper and lower control limits for the $\bar{X}$-chart is 2.8. So, $6\sigma_{\bar{x}} = 2.8$ or $\sigma_{\bar{x}} = 0.467$. An estimate of the standard deviation of the waiting time is $\hat{\sigma} = 0.467\sqrt{4} = 0.934$. Using Minitab, with a subgroup size of one and a standard deviation of 0.467 (since we are monitoring the averages), a moving-average chart using

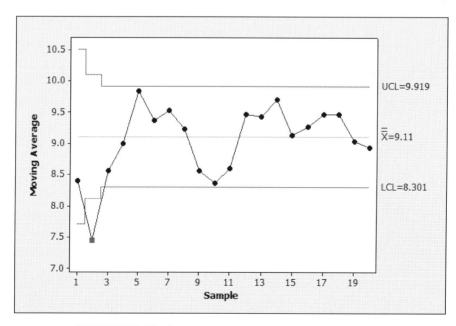

FIGURE 7-5. Moving average chart of average waiting time

112

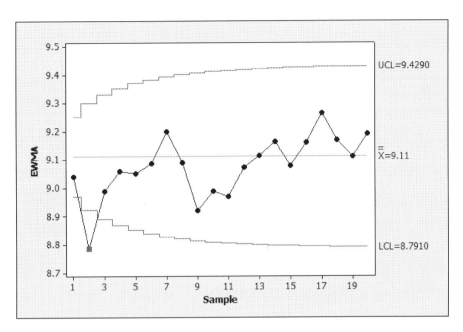

FIGURE 7-6. Geometric moving average chart of average waiting time

a span of 3 is shown in Figure 7-5. Sample number 2 is out-of-control. However, it is on the desirable side as far as waiting time is concerned.

7-34.   For the data in Problem 7-33, an estimate of the standard deviation of waiting time is $\hat{\sigma} = 0.934$. A geometric moving average chart for a weighting factor of 0.1 is shown in Figure 7-6. The chart is constructed using a value for the standard deviation (of the averages since that is being monitored here) of 0.467. Sample number 2 plots below the LCL and is out-of-control, which was also the case in Problem 7-33. The plotted points in this problem are further removed from the control limits compared to those in Problem 7-33.

7-35.   An individuals and moving-range chart (with a window of 2) is constructed using Minitab and is shown in Figure 7-7. Even though none of the observations are outside the control limits on both charts, on the individuals chart a somewhat cyclic behavior is observed.

7-36.   a)   An individuals and moving-range chart is shown in Figure 7-8. All the plotted points are within the control limits on both charts.

        b)   A moving-average chart is shown in Figure 7-9. It displays a pattern similar to that of the individuals chart.

        c)   An exponentially weighted moving-average chart is shown in Figure 7-10. Here, a visible downward trend is observed for samples 11 through 40. Further, observation number 4 is slightly above the UCL.

113

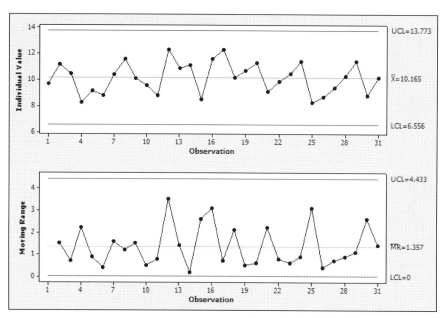

FIGURE 7-7. Individual and moving range chart of processing times before improvement

d) A cumulative sum chart, with a target value of 30 seconds, is shown in Figure 7-11. Note that observation number 11, with a cumulative sum of 15.42 is the first observation to plot above the UCL. Observation number 12, with a cumulative sum of 17.69, and observation number 13, with a cumulative sum of 13.76, also plot above the UCL. Hence, the cumulative sum chart is sensitive enough to detect an upward shift in the process.

7-37. From the given data, the following summary statistics are obtained: $\sum \overline{X} = 2197.5$,

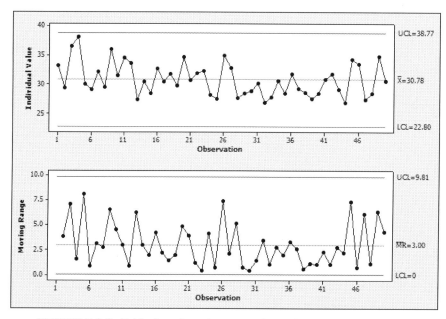

FIGURE 7-8. Individuals and moving range chart of call waiting time

114

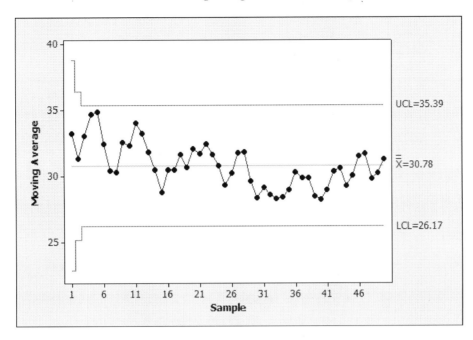

FIGURE 7-9. Moving average chart of call waiting time

$\sum R = 49.9$, $\sum i = 210$, $\sum Xi = 23431.8$, $\sum i^2 = 2870$. The parameters of the fitted centerline are:

$$a = \frac{(2197.5)(2870) - (23431.8)(210)}{20(2870) - (210)^2} = 104.222$$

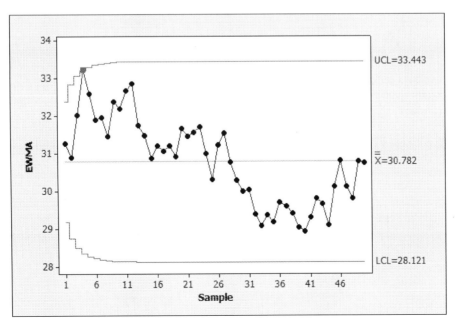

FIGURE 7-10. EWMA chart of call waiting time

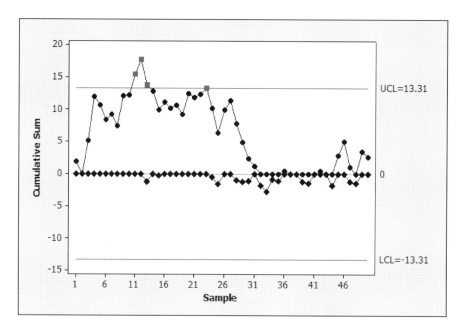

FIGURE 7-11. CUSUM chart of call waiting time

$$b = \frac{20(23431.8) - (2197.5)(210)}{20(2870) - (210)^2} = 0.538.$$

The centerline is given by: $C = 104.222 + 0.538i$.

Average of the ranges is $\bar{R} = 49.9/20 = 2.495$. The control limits for the trend chart are:

$$UCL = (a + A_2\bar{R}) + bi = [104.22 + 0.577(2.495)] + 0.538i = 105.660 + 0.538i.$$

$$LCL = (a - A_2\bar{R}) + bi = [104.22 - 0.577(2.495)] + 0.538i = 102.780 + 0.538i.$$

Several of the sample averages plot outside the control limits as shown in Figure 7-12. The die should be changed when the centerline reaches a value that is 3σ below the upper specification limit. An estimate of the process standard deviation is $\bar{R}/d_2 = 2.495/2.326 = 1.073$. Thus, the maximum value that the centerline should be allowed to reach is $118 - 3(1.073) = 114.781$, at which point the die should be changed.

7-38. From the given data, $\bar{R} = 51.6/25 = 2.064$. The control limits for the R-chart are: UCL = $D_4\bar{R} = 2.282(2.064) = 4.710$, LCL = $D_3\bar{R} = 0(2.064) = 0$. All of the range values are within the control limits. With the variability in control, we now calculate the modified control limits.

For $\alpha = 0.05$, $Z_a = 1.645$, and for $\delta = 0.015$, $Z_\delta = 2.17$. An estimate of the process standard deviation is $\hat{\sigma} = \bar{R}/d_2 = 2.064/2.059 = 1.002$. The modified control limits are:

$$LCL = 18 + \left(2.17 - \frac{1.645}{\sqrt{4}}\right) 1.002 = 19.350$$

$$UCL = 35 - \left(2.17 - \frac{1.645}{\sqrt{4}}\right) 1.002 = 33.650.$$

Average for sample number 7 falls below the lower control limit.

7-39.   From the given data, $\bar{R} = 1.99$, with the control limits for the R-chart being UCL = 4.207, LCL = 0, and all of the range values being in control. An estimate of the process standard deviation is $\hat{\sigma} = \bar{R}/d_2 = 1.99/2.326 = 0.856$. For $\alpha = 0.025$, $Z_\alpha = 1.96$, and for $\delta = 0.005$, $Z_\delta = 2.75$. The modified control limits are:

$$LCL = 12 + \left(2.575 - \frac{1.96}{\sqrt{5}}\right) 0.856 = 13.454.$$

$$UCL = 33 - \left(2.575 - \frac{1.96}{\sqrt{5}}\right) 0.856 = 31.546.$$

All of the sample averages are within these modified control limits. They are well below the upper control limit and are closer to the lower control limit.

7-40.   For the data in Problem 7-38, $\bar{R} = 2.064$ and $\hat{\sigma} = 1.002$. Here $\gamma = 0.04$, and $1 - \beta = 0.90$, which yields $Z_\gamma = 1.75$ and $Z_\beta = 1.28$. The acceptance control chart limits are:

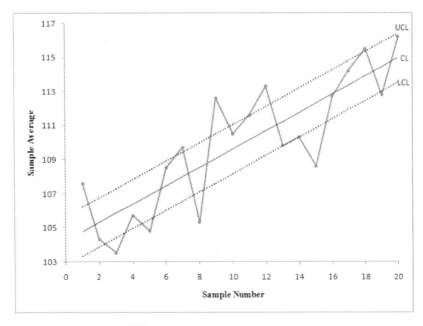

FIGURE 7-12. Trend chart for diameter

$$LCL = 18 + \left(1.75 + \frac{1.28}{\sqrt{4}}\right) 1.002 = 20.395.$$

$$UCL = 35 - \left(1.75 + \frac{1.28}{\sqrt{4}}\right) 1.002 = 32.605.$$

Averages for sample numbers 2, 4, 7, 9, 10, 16, and 17 fall below the lower control limit.

7-41. For the data in Example 7-14, $\bar{R} = 1.99$ and $\hat{\sigma} = 0.856$. Here $\gamma = 0.02$ and $1-\beta = 0.98$, which yields $Z_\gamma = 2.054$ and $Z_\beta = 2.054$. The acceptance control chart limits are:

$$LCL = 12 + \left(2.054 + \frac{2.054}{\sqrt{5}}\right) 0.856 = 14.544.$$

$$UCL = 33 - \left(2.054 + \frac{2.054}{\sqrt{5}}\right) 0.856 = 30.455.$$

All of the sample averages are within these acceptance control chart limits.

7-42. For the given data, for each sample, the sample means for tensile strength and diameter, the sample variances for each of the two characteristics, the sample covariance, and Hotelling's T^2 statistic are shown in Table 7-1. Notation-wise, m = 20, n = 4, p = 2.

The upper control limit for the T^2 chart for an overall type I error probability of 0.01 is:

$$UCL = \left[\frac{(20)(4)(2) - (20)(2) - (4)(2) + 2}{(20)(4) - 20 - 2 + 1}\right] F_{0.01, 2, 59}$$

$$= (1.932)(4.988) = 9.637.$$

A multivariate T^2 control chart is shown in Figure 7-13. Sample number 7 has a T^2 value of 18.263, that plots above the UCL and is out of control. Let us calculate the individual control limits for sample number 7 for each of the two quality characteristics. For tensile strength, the control limits are:

$$71.262 \pm t_{.0025,60}\sqrt{18.904}\sqrt{\frac{19}{80}},$$

$$= 71.262 \pm (3.0175)(2.1189) = 71.262 \pm 6.394 = (64.868, 77.656).$$

118

TABLE 7-1. Calculations on Hotelling's T^2 for Tensile Strength and Diameter

Sample number j	Sample means		Sample variances		Sample covariance	Hotelling's T^2
	Tensile strength $\overline{X}_{1j}$	Diameter $\overline{X}_{2j}$	s_{1j}^2	s_{2j}^2	s_{12j}	
1	69.000	17.250	6.667	4.917	5.000	8.835
2	70.000	19.000	50.000	4.667	-13.333	0.586
3	67.500	17.750	8.333	4.250	-4.167	8.312
4	70.000	17.750	17.667	4.917	-2.500	4.173
5	72.250	21.000	2.917	2.667	1.333	5.143
6	73.250	19.500	0.917	1.667	-1.167	1.033
7	64.000	22.000	3.333	2.667	2.000	18.263
8	75.000	21.000	54.000	1.333	-6.000	9.160
9	70.500	17.500	25.000	1.000	1.000	5.528
10	68.000	18.000	2.000	0.667	-0.333	6.033
11	73.500	21.000	25.667	6.667	4.667	6.555
12	65.750	20.750	152.250	0.917	-10.750	8.052
13	76.000	19.250	8.000	2.917	-4.000	4.910
14	74.500	18.250	1.667	1.583	1.167	3.210
15	72.000	17.250	3.333	0.917	0.667	6.319
16	68.500	18.250	1.667	2.917	1.833	4.120
17	75.000	21.000	4.667	3.333	-2.333	9.160
18	73.500	20.750	1.667	2.917	-1.167	8.760
19	73.500	19.250	1.667	0.917	-0.833	1.084
20	71.000	19.500	6.667	1.667	-0.667	0.065
Means	$\overline{\overline{X}}_1 = 71.262$	$\overline{\overline{X}}_2 = 19.300$	$s_1^2 = 18.904$	$s_2^2 = 2.675$	$s_{12} = -1.479$	

The sample average tensile strength of 64 falls slightly below the lower control limit. So, one might look for special causes that deal with tensile strength for sample number 7.

Similarly, for the diameter, the control limits are:

$$19.300 \pm (3.0175)\sqrt{2.675}\sqrt{\frac{19}{80}},$$

$$= 19.300 \pm 2.405 = (16.895, 21.705).$$

The sample average diameter of 22.000 for sample number 7, is slightly above the upper control limit. Thus, one might also look for special causes related to diameter.

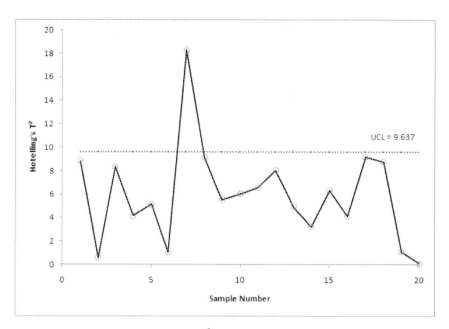

FIGURE 7-13. Multivariate T^2 chart for strength and diameter

7-43. A standardized control chart for individuals (Z) and a moving range (MR) chart of the project completion times is constructed. For each level of project complexity, the mean and standard deviation of completion time are estimated. Using Minitab, a Z-MR-chart is shown in Figure 7-14. The plotted values are well within the control limits on both charts.

7-44. a) Using Minitab, a T^2 chart is constructed for the process parameters temperature, pressure, proportion of catalyst, and acidity. The values of T^2 for observation number 4 (value = 16.11) and observation number 8 (value = 18.29) fall above the

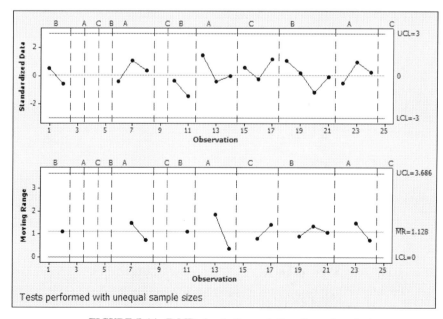

FIGURE 7-14. Z-MR chart of completion time of projects

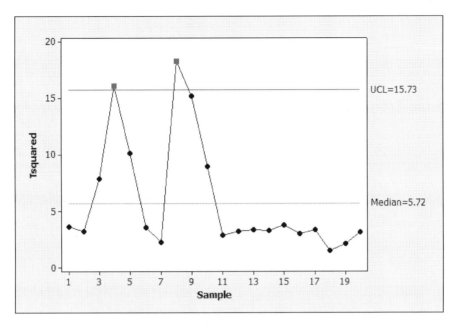

FIGURE 7-15. T^2 chart of process parameters

upper control limit indicating an out-of-control situation. Figure 7-15 shows the multivariate control chart. Minitab also outputs the following p-values associated with the characteristics for the two observations greater than the UCL. For observation 4: Temperature (p-value = 0.0072); Pressure (p-value = 0.0121). For observation 8: pressure (p-value = 0.0000); Proportion of catalyst (p-value = 0.0005); Acidity (p-value = 0.0012). Hence, these above mentioned characteristics should be investigated.

b) A generalized variance chart, using Minitab, is shown in Figure 7-16. All of the plotted values are well within the control limits.

7-45. a) A Hotelling's T^2 chart is constructed on patient characteristics before the drug is administered. Figure 7-17 shows the T^2 chart and the generalized variance chart using Minitab. All points on both charts are within the control limits.

b) An individuals and moving-range chart for blood glucose level before administration of the drug is constructed and shown in Figure 7-18. Note that on the individuals chart, for observation 15, the blood glucose level exceeds theUCL, while the corresponding moving-range value also exceeds the UCL on the MR-chart. If only blood glucose level were to be considered, observation 15 has a value that is above what is expected. However, if all patient characteristics are considered, the T^2 value for this observation is not considered unusual. Thus, the conclusions from parts a) and b) differ.

121

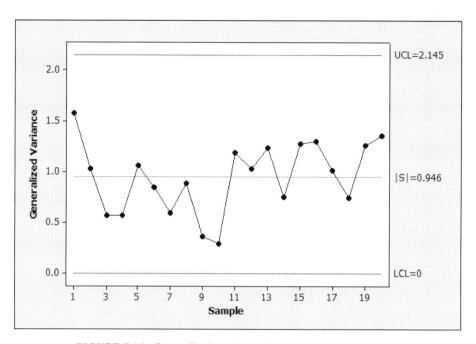

FIGURE 7-16. Generalized variance chart on process parameters

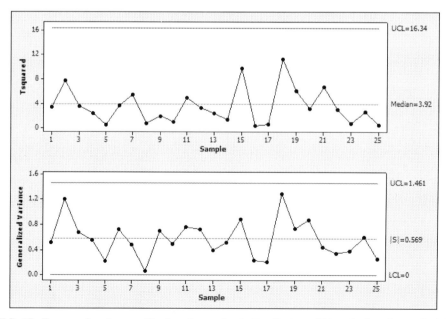

FIGURE 7- 17. Tsquared and generalized variance chart of patient conditions before drug administration

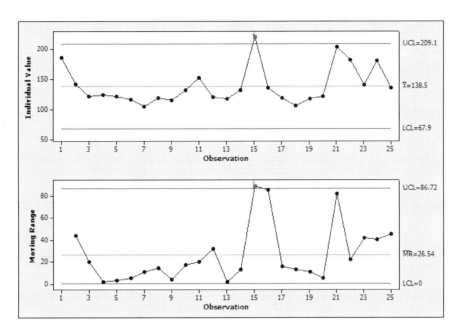

FIGURE 7-18. I and MR chart for glucose levels before drug administration

CHAPTER 8

CONTROL CHARTS FOR ATTRIBUTES

8-1.   a)   Examples of nonconformities are errors in customer monthly statements or errors in a loan processing application. On the other hand, rather than count errors, if we define a customer statement as either error-free or not, or a loan processing application as either error-free or not, it would be an example of a nonconforming item.

        b)   Examples of nonconformities include number of medication errors or errors in laboratory analysis. Nonconforming items include whether a patient is not satisfied or whether a hospital bed is not available.

        c)   Example of nonconformities includes number of defective solders in a circuit board, while a nonconforming item could be the circuit board being not defect-free.

        d)   An example of nonconformity is the number of unsubstantiated references in a legal document, while a nonconforming item could be a case that is lost in court.

        e)   An example of nonconformity is the number of errors in allocating funds, while a nonconforming item could be the improper distribution of a certain donor's gift.

8-2   Certain characteristics are measured as attributes, for example, the performance of a staff member. The number of control charts required could be less when using an attribute chart. For example, several characteristics could be lumped together such that when all criteria are satisfied, the item is classified as acceptable. Further, attribute charts can be used at various levels in the organization whereas variables chart are used at the lowest levels (individual person or operator). One disadvantage is that attribute charts do not provide as much information as variables charts. Also, the response time to detect a shift in the process is usually slower and the sample sizes required, for similar levels of protection, are larger than that for variables charts.

8-3.   For a p-chart, the choice of an appropriate sample size is critical. The sample size must be such that it must allow for the possibility of occurrences of nonconforming items in the sample. As an example, if a process has a nonconformance rate of 1%, a sample size of 400 or 500 is necessary.

8-4.   A p-chart for the proportion nonconforming should be used and should be at the overall organization level. Thus, if the CEO has responsibility for 5 plants, the p-chart should measure product output quality over these plants. Hopefully, through such monitoring, one could obtain an indication of a specific plant(s) which does not perform up to expectations.

8-5.   A change in the sample size does not affect the centerline on a p-chart. The control limits are drawn closer to the centerline with an increase in the sample size.

8-6.   Since the proportion nonconforming values are normalized in a standardized p-chart, the control limits remain constant, even though the subgroup size may vary. These limits are at $\pm$ 3. Also, the tests for detection of out-of-control patterns using runs are easier to apply than the regular p-chart where the subgroup size changes.

8-7. The assumptions for a p-chart are those associated with the binomial distribution. This implies that the probability of occurrence of a nonconforming item remains constant for each item, and the items are assumed to be independent of each other (in terms of being nonconforming). The assumptions for a c-chart, that deals with the number of nonconformities, are those associated with the Poisson distribution. The opportunity for occurrence of nonconformities could be large, but the average number of nonconformities per unit must be small. Also, the occurrences of nonconformities must be independent of each other. Further, the chance of occurrence of nonconformities should be the same from sample to sample.

8-8. When the p-chart is constructed based on data collected from the process, it is quite possible for the process to be in control and still not meet desirable standards. For example, if the desired standards are very stringent, say a 0.001% nonconformance rate, the current process may not be able to meet these standards without major changes. Remedial actions would involve systemic changes in the process that reduce proportion nonconforming. It could be through change in equipment, training of personnel, or scrutiny in selection of vendors. The detection could take place if the centerline and control limits are calculated based on the desirable standard and data from the current process is plotted on that chart.

8-9. Customer satisfaction or "acceptance" of the product or service may influence the p-chart. A survey of the customers may indicate their needs, based on which management of the organization could design the product/service appropriately. Customer feedback becomes crucial for determining the ultimate acceptance of the product/service. In a total quality system approach, the customer is part of the extended process. Determining customer expectations provides valuable information to product and process design.

8-10. If the control limits are expanded further out from the centerline, the chance of a false alarm (type I error) will decrease, implying that the ARL, for an in-control process, will increase. The operating characteristic curve represents the probability of failing to detect a process change, when a process change has taken place, which is the probability of a type II error. So, in this situation with the expanded control limits, the ARL to detect a change, for an out-of-control process, will also increase.

8-11. In monitoring the number of nonconformities, when the sample size changes from sample to sample, a u-chart is used to monitor the number of nonconformities per unit.

8-12. For highly conforming processes, the occurrence of nonconformities or nonconforming items is very rare. Hence, extremely large sample sizes will be necessary to ensure the observation of such, in order to construct a p-chart or a c-chart. Often times, this may not be feasible. An alternative could be to observe the time or number of items to observe a nonconformity. Further, when the proportion nonconforming is very small, the normal distribution is not a good approximation to the binomial distribution.

Also the p- or c-chart may show an increased false alarm rate for highly conforming processes and may also fail to detect a process change, when one takes place. When the proportion nonconforming is very small, LCL may be less than 0. So, process improvement cannot be detected.

8-13. Three-sigma limits are based on the assumption of normality of distribution of the statistic being monitored. So, when the distribution of the statistic cannot be reasonably approximated by the normal distribution, probability limits, that are based on the actual distribution of the statistic, should be used. For example, the time to observe a defect could have an exponential distribution. So, the exponential distribution should be used to find say the lower and upper control limits (say at 0.13% and 99.87%) of the distribution. These limits may not be symmetric about the centerline.

8-14. An attribute chart could monitor the proportion of time customer due dates are met through a p-chart. In this case, the assumptions associated with a binomial distribution would have to be met. It means that the chance of an order meeting the due date should remain constant across all orders and the outcome of one order in terms of meeting the due date should be independent of other orders. In terms of a variable chart, we could monitor the number of days that the due date is missed by for each order, through an individuals and moving-range chart. For this situation, the assumption of normality of the distribution of the number of days that the due date is missed by is appropriate. We also assume that the orders are independent or each other so that the delay in one order has no influence on any other order.

8-15. When defects or nonconformities have different degrees of severity, a U-chart representing demerits per unit is used. The degree of severity of a defect could be influenced by the corresponding user of the product/service. Thus, what one customer perceives as "poor" service in a restaurant could be different from another. Therefore, based on the context of use of the product/service, appropriate severity ratings should be established.

8-16. a) If fixed segments of the highway (say, 5 kilometers) are randomly selected for each sample, a c-chart is appropriate. If the chosen segment varies from sample to sample, a u-chart that monitors the number of potholes/kilometer could be used.

b) p-chart.

c) U-chart.

d) p-chart that monitors the proportion of errors.

e) p-chart to control the proportion of claims filed. If the number of insured persons remains constant from month to month, an np-chart could be used.

f) U-chart.

g)   u-chart.

h)   p-chart.

i)   u-chart.

j)   c-chart (assuming the population size remains approximately constant during the period), otherwise a u-chart.

k)   p-chart.

l)   For a given type of aircraft, a c-chart. Otherwise, if different aircrafts have a different number of welds, a u-chart.

m)   p-chart.

n)   p-chart.

o)   c-chart that monitors the number of calls per unit time period.

p)   p-chart.

q)   c-chart.

r)   U-chart.

s)   c-chart (assuming the number of vehicles remains approximately constant), otherwise u-chart.

t)   p-chart.

8-17. A p-chart is constructed for the employee's proportion of processing errors. The centerline is $\bar{p} = 197/9100 = 0.0216$. The control limits for the first 16 samples are given by:

$$0.0216 \pm 3 \sqrt{\frac{(0.0216)(0.9784)}{400}}$$

$$= 0.0216 \pm 0.0218 = (0, 0.0434).$$

For the last 9 samples, the control limits are given by:

$$0.0216 \pm 3 \sqrt{\frac{(0.0216)(0.9784)}{300}}$$

$= 0.0216 \pm 0.0252 = (0, 0.0468).$

Sample 14, with a proportion nonconforming of 0.045 plots above the UCL, as shown in Figure 8-1. Special causes need to be investigated for this sample so that appropriate remedial actions might be identified. If observation number 14 is deleted, the revised centerline is $\bar{p} = 179/8700 = 0.0206$. A revised p-chart shows all points within the control limits and a stable pattern.

The average proportion of errors expected from this employee would be 0.0206 or 2.06 %. It would be unusual to expect error-free performance from this employee, even though the employee might occasionally achieve this level.

8-18. Centerline $\bar{p} = 83/2000 = 0.0415$. Trial control limits are:

$$0.0415 \pm 3 \sqrt{\frac{(0.0415)(0.9585)}{100}} = 0.0415 \pm 0.0598 = (0, 0.1013).$$

Sample numbers 10 and 16 are above the UCL. Deleting these two, the revised centerline is $\bar{p} = (83-11-12)/1800 = 60/1800 = 0.0333$. The revised control limits are:

$$0.0333 \pm 3 \sqrt{\frac{(0.0333)(0.9667)}{100}} = 0.0333 \pm 0.0538 = (0, 0.0871).$$

8-19. Control limits based on the standard of 3 percent are:

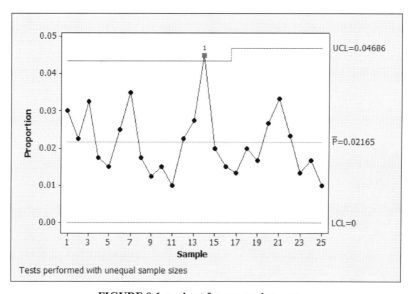

FIGURE 8-1. p-chart for processing errors

130

$$0.03 \pm 3 \sqrt{\frac{(0.03)(0.97)}{100}} = 0.03 \pm 0.051 = (0, 0.081).$$

The current revised process average of 3.33% is slightly above the standard of 3%. All of the observations would be within the control limits determined by the standard. The process could meet this standard.

If the standard is set at 2 percent, the control limits would be:

$$0.02 \pm 3 \sqrt{\frac{(0.02)(0.98)}{100}} = 0.02 \pm 0.042 = (0, 0.062).$$

The process average is above this standard of 2%. However, the process variability is small enough such that all of the observations (upon revision) are within these control limits. The process should meet this standard and management should investigate means of reducing the process average.

8-20.   A p-chart for the proportion of errors is constructed for the department. The centerline is:

$$\bar{p} = 170/9900 = 0.0172.$$

The control limits are as follows:

For observations 1-6:

$$0.0172 \pm 3 \sqrt{\frac{(0.0172)(1 - 0.0172)}{400}}$$

$$= 0.0172 \pm 0.0195 = (0, 0.0367).$$

For observations 7-16:

$$0.0172 \pm 3 \sqrt{\frac{(0.0172)(1 - 0.0172)}{300}}$$

$$= 0.0172 \pm 0.0225 = (0, 0.0397).$$

For observations 17-25:

$$0.0172 \pm 3 \sqrt{\frac{(0.0172)(1 - 0.0172)}{500}}$$

$$= 0.0172 \pm 0.0174 = (0, 0.0346).$$

Observation number 10 is above the UCL. Assuming special causes and deleting this observation, the revised centerline is:

$$\bar{p} = 155/9600 = 0.0161.$$

The revised control limits are as follows:

For observations 1-6:

$$0.0161 \pm 3 \sqrt{\frac{(0.0161)(1 - 0.0161)}{400}}$$

$$= 0.0161 \pm 0.0189 = (0, 0.0350).$$

For observations 7-16:

$$0.0161 \pm 3 \sqrt{\frac{(0.0161)(1 - 0.0161)}{300}}$$

$$= 0.0161 \pm 0.0218 = (0, 0.0379).$$

For observations 17-25:

$$0.0161 \pm 3 \sqrt{\frac{(0.0161)(1 - 0.0161)}{500}}$$

$$= 0.0161 \pm 0.0169 = (0, 0.0330).$$

All of the observations are now within the control limits. The average proportion of errors is 0.0161. For a target value of 0, the number of standard deviations that the average is from 0 when the number sampled is 300 is $0.0161/0.0073 = 2.216$. So it may not be feasible to expect error-free performance from this department on average, even though, on occasions there could be zero errors.

8-21. The center line is $\bar{p} = 181/3460 = 0.0523$. The control limits for each subgroup are obtained from:

$$0.0523 \pm 3 \sqrt{\frac{(0.0523)(0.9477)}{n_i}} = 0.0523 \pm \frac{0.6679}{\sqrt{n_i}}.$$

Table 8-1 shows the fraction nonconforming and control limits for each subgroup.

TABLE 8-1. Control Limits on Proportion Nonconforming

Sample number	Fraction nonconforming	LCL	UCL	Sample number	Fraction nonconforming	LCL	UCL
1	0.0375	0	0.1270	14	0.0444	0	0.1227
2	0.0500	0	0.1133	15	0.0312	0	0.1051
3	0.0667	0	0.1385	16	0.0130	0.0083	0.0963
4	0.0333	0	0.1068	17	0.0600	0.0051	0.0995
5	0.0571	0	0.1087	18	0.0533	0	0.1068
6	0.0667	0	0.1068	19	0.0286	0.0062	0.0984
7	0.0437	0	0.1051	20	0.0210	0.0038	0.1007
8	0.0667	0	0.1227	21	0.0562	0	0.1051
9	0.0500	0	0.1191	22	0.0800	0	0.1191
10	0.0750	0	0.1051	23	0.1200	0	0.1191
11	0.0727	0	0.1160	24	0.0778	0	0.1227
12	0.5000	0	0.1227	25	0.0625	0	0.1051
13	0.0700	0.0051	0.0995				

Figure 8-2 shows the p-chart. Sample number 23 with a fraction nonconforming of 0.12 is above the UCL and out of control. The revised centerline is $\bar{p} = (181 - 12)/3360 = 169/3360 = 0.0503$. The revised control limits are given by:

$$0.0503 \pm 3 \sqrt{\frac{(0.0503)(0.9497)}{n_i}} = 0.0503 \pm \frac{0.6557}{\sqrt{n_i}}.$$

8-22. The process average was calculated as $\bar{p} = 0.0523$. The standard deviation of the sample fraction nonconforming for the ith sample is given by:

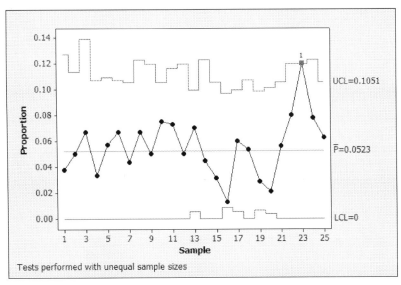

FIGURE 8-2. p-chart for nonconforming items

133

$$\sigma_{\hat{p}} = \sqrt{\frac{(0.0523)(0.9477)}{n_i}} = \frac{0.2226}{\sqrt{n_i}}.$$

The standardized values are shown in Table 8-2. Since the control limits are at 3 and -3, sample number 23 with a standardized value of 3.041 is above the UCL and out of control. The same conclusions, as drawn previously in Problem 8-21, are drawn here.

8-23. A p-chart for the proportion of medication errors is constructed as shown in Figure 8-3. Note that the sample size used is 1000 for each sample. For sample 15, the proportion of medication errors is 0.042 and exceeds the upper control limits.

Assuming special causes are identified and remedial actions are taken, we construct the revised limits. The revised centerline is $\overline{p} = 50.1/24 = 0.0209$. The revised control limits are:

$$0.0209 \pm 3 \sqrt{\frac{(0.0209)(0.9791)}{1000}}$$

$$= 0.0209 \pm 0.0136 = (0.0073, 0.0345).$$

TABLE 8-2. Standardized Value of Fraction Nonconforming

Sample number	Fraction non-conforming	Standard deviation	Standard-ized value	Sample number	Fraction non-conforming	Standard deviation	Standard-ized value
1	0.0375	0.0249	-0.595	14	0.0444	0.0235	-0.337
2	0.0500	0.0203	-0.113	15	0.0312	0.0176	-1.199
3	0.0667	0.0287	0.501	16	0.0130	0.0147	-2.677
4	0.0333	0.0182	-1.045	17	0.0600	0.0157	0.489
5	0.0571	0.0188	0.255	18	0.0533	0.0182	0.055
6	0.0667	0.0182	0.792	19	0.0286	0.0154	-1.543
7	0.0437	0.0176	-0.489	20	0.0210	0.0161	-1.938
8	0.0667	0.0235	0.614	21	0.0562	0.0176	0.222
9	0.0500	0.0223	-0.103	22	0.0800	0.0223	1.244
10	0.0750	0.0176	1.290	23	0.1200	0.0223	3.041
11	0.0727	0.0212	0.961	24	0.0778	0.0235	1.087
12	0.0500	0.0223	-0.103	25	0.0625	0.0176	0.560
13	0.0700	0.0157	1.124				

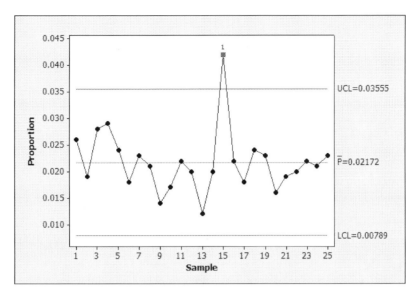

FIGURE 8-3. p chart for medication errors

The remaining samples are now in control. The mean of the proportion of medication errors is 0.0209 or 2.09%. This is much higher (4.61 standard deviations) than the target value of zero, which represents error-free performance. Hence, it is unreasonable to expect error-free performance from the current system. Fundamental changes must be made by the administration. Areas where such changes are to be made can be identified through Pareto analysis and subsequent cause-and-effect analysis of the main reasons behind medication errors.

8-24.   a)   A p-chart is constructed for the proportion of C-sections and is shown in Figure 8-4. The process is in control.

   b)   While the average C-section rate is higher for the last 6 months relative to the previous 6 months, it is not statistically significant.

   c)   If no changes are made in current obstetrics practices, the C-section rate is predicted to be about 15.31%.

   d)   The difference between 15.31% and the benchmark value of 10% is approximately 3.01 in standard deviation units. This is found as follows:

$$\sigma_{\hat{p}} \simeq \frac{0.2058 - 0.1531}{3} = 0.0176.$$

$$Z = \frac{0.1531 - 0.10}{0.0176} = 3.01.$$

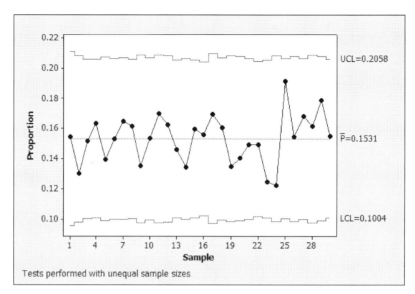

FIGURE 8-4. p-chart for C-sections

Hence, if no changes are made to the current system, it does not seem feasible to achieve the benchmark value of 10%.

8-25. The centerline for the np-chart is 83/20 = 4.15, with the control limits being:

$$4.15 \pm 3\sqrt{(4.15(1-4.15/100)} = 4.15 \pm 5.983 = (0, 10.133).$$

Sample numbers 10 and 16 plot above the UCL. Deleting these two, the revised centerline is (83-23)/18 = 3.333. The revised control limits are:

$$3.333 \pm 3\sqrt{3.333(1-3.333/100)} = 3.333 \pm 5.385 = (0, 8.718).$$

8-26. A c-chart for the number of processing errors is shown in Figure 8-5. Sample 9, with 11 errors, plots above the UCL.

Assuming that special causes are identified for sample 9, and remedial actions are taken, we revise the control chart. The revised centerline is:

$$\bar{c} = \frac{90}{24} = 3.750.$$

The revised control limits are:

$$3.750 \pm 3\sqrt{3.750} = (0, 9.559).$$

136

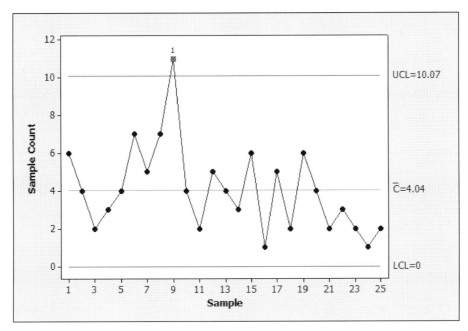

FIGURE 8-5 c-chart for processing errors

Let us examine the impact of the new purchase order form. Even though the number of processing errors did increase by 1 for the last sample, a downward trend is observed. The preliminary indication is that the new form is an improvement over the old one. However, further samples are needed to confirm this inference since only five samples from the new form have been observed.

The revised centerline and control limits reflect the capability of the process. An estimate of the standard deviation of the number of processing errors per 100 purchase orders is $\sqrt{3.750} = 1.936$. The revised centerline of 3.75 is removed from the goal value of 0 in standard deviation units as:

$$Z = \frac{3.75 - 0}{1.936} = 1.937 \approx 1.94.$$

It is therefore possible, but not probable, to achieve zero errors on a consistent basis. Perhaps, as more data is obtained using the new form, a better estimate of the chances of meeting the goal can be derived.

8-27. a) A c-chart is constructed for the number of dietary errors per 100 trays. The centerline is:

$$\bar{c} = \frac{181}{25} = 7.24.$$

The control limits are:

$7.24 \pm 3 \sqrt{7.24} = 7.24 \pm 8.072 = (0, 15.312).$

The c-chart is shown in Figure 8-6.

Sample number 7 is above the UCL. Assuming special causes have been identified and remedial actions taken, the revised centerline is:

$$\overline{c} = \frac{181 - 16}{24} = 6.875.$$

The revised control limits are:

$6.875 \pm 3 \sqrt{6.875} = 6.875 \pm 7.866 = (0, 14.741).$

b)   If no changes are made in the process, we would expect, on average, 6.875 dietary errors per 100 trays.

c)   If the process average is 6.875 errors/100 trays, using the Poisson distribution, the probability of 2 or fewer dietary errors is:

$P(X \leq 2) = 0.0326.$

Thus, without changes in the process, it would not be feasible to achieve this level of capability.

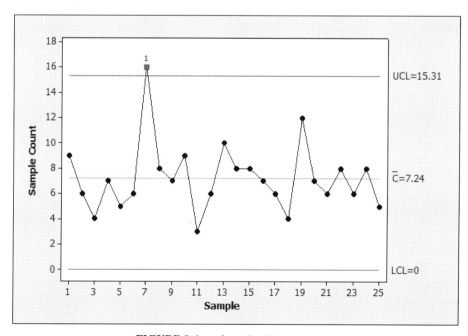

FIGURE 8-6. c-chart for dietary errors

8-28. The centerline for a c-chart is 80/30 = 2.667, with the control limits being:

$$2.667 \pm 3 \sqrt{2.667} = 2.667 \pm 4.899 = (0, 7.566).$$

Using the specified goal, the centerline on a c-chart is 0.5 blemishes per 100 square meter, with the control limits being:

$$0.5 \pm 3 \sqrt{0.5} = 0.5 \pm 2.121 = (0, 2.621).$$

The process average is at 2.667/3 = 0.889 blemishes per 100 square meter, with the process standard deviation being $\sqrt{0.889}$ = 0.943. The process average exceeds the goal value of 0.5. Let us calculate the probability of the number of blemishes not exceeding the goal UCL of 2.621, given that the process average is 0.889. Using the Poisson distribution, we have $P[X < 2.621 \mid \lambda = 0.889] = P[X \leq 2 \mid \lambda = 0.889] = 0.9389$. So, about 6.11% of the time the process will be deemed to be out of control, making the process not totally capable.

8-29. The centerline on a u-chart for the number of imperfections per square meter is 189/4850 = 0.039. The control limits are given by:

$$0.039 \pm 3 \sqrt{\frac{0.039}{n_i}}.$$

Figure 8-7 shows the u-chart for the number of imperfections per square meter.

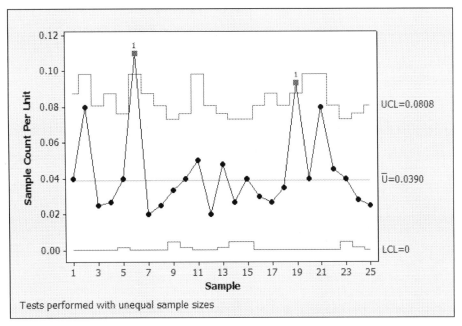

FIGURE 8-7 u-chart of imperfections

139

Sample numbers 6 and 19 plot above the UCL and are out of control. Deleting these two samples, the revised centerline is (189-11-14)/(4850-100-150) = 0.0356. The revised control limits are given by:

$$0.0356 \pm 3 \sqrt{\frac{0.0356}{n_i}}.$$

The revised control limits are shown in Table 8-3.

8-30. The calculation of the subgroup size, u_i values, centerline, and control limits would be affected, since the unit is now 100 square meters. However, the decisions would not change. For example, the trial centerline would be 189/48.5 = 3.897 imperfections per 100 square meters. The control limits would be given by:

$$3.897 \pm 3 \sqrt{\frac{3.897}{n_i}},$$

where n_i is the subgroup size in units of 100 square meters. As an example, for sample number 1, the control limits would be:

$$3.897 \pm 3 \sqrt{\frac{3.897}{1.5}} = (0, 8.732),$$

with the u_i value being 6/1.5 = 4 imperfections per 100 square meters. The relative shape of the plot will be the same as in Problem 8-29 and there will be no change in decision making. The conclusions that we may draw are that the units selected for measurement (be it in square meters or 100 square meters) do not influence the decisions that are made from control charts. The user may select one that is convenient for the appropriate setting and the charts will be scaled accordingly.

8-31. A u-chart for the number of medication errors per order filled is constructed. The centerline is obtained as:

$$\bar{u} = \frac{312}{29560} = 0.01055.$$

TABLE 8-3. Revised Control Limits for the u-Chart

Subgroup size	Sample numbers	LCL	UCL	Subgroup size	Sample numbers	LCL	UCL
100	2,6,11,20,21	0	0.0922	250	5,10,13	0	0.0714
150	1,4,7,17,19	0	0.0818	300	9,14,15,23	0.0029	0.0683
200	3,8,12,16,18, 22,25	0	0.0756				

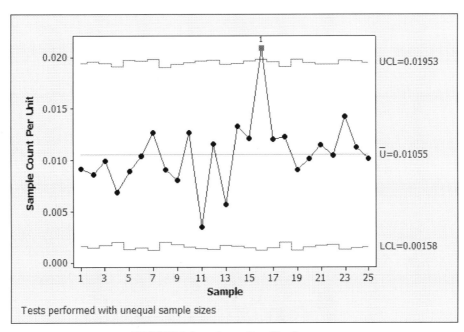

FIGURE 8-8. u-chart of medication errors

The control limits are given by:

$$0.01055 \pm 3 \sqrt{\frac{0.01055}{n_i}}.$$

Figure 8-8 shows the u-chart.

Note that sample 16 with a u-value of 0.0291 plots above the UCL. Assuming that special causes have been identified and remedial actions taken, the revised centerline is obtained as:

$$\bar{u} = \frac{312 - 23}{28360} = 0.01019.$$

A revised control chart may be constructed using the remaining values.

8-32. The average number of serious defects per unit is $\bar{u}_1 = 8/[(5)(25)] = 0.064$. Similarly, the average number of major defects per unit is $\bar{u}_2 = 75/125 = 0.6$, and the average number of minor defects per unit is $\bar{u}_3 = 158/125 = 1.264$. The centerline of a demerits per unit chart is:

$$\overline{U} = 50(0.064) + 10(0.6) + 1(1.264) = 10.464.$$

The standard deviation of U is

$$\hat{\sigma}_U = \sqrt{\frac{50^2(0.064) + 10^2(0.6) + 1.264}{5}} = 6.652.$$

Control limits are found as:

$10.464 \pm 3(6.652) = 10.464 \pm 19.956 = (0, 30.42)$.

The demerits per unit for each sample are shown in Table 8-4.

Sample number 10 plots above the UCL. The revised averages are:

$\bar{u}_1 = 6/120 = 0.05$; $\bar{u}_2 = 69/120 = 0.575$; $\bar{u}_3 = 152/120 = 1.267$.

The revised centerline is:

$\bar{U} = 50(0.05) + 10(0.575) + 1(1.267) = 9.517$.

The revised standard deviation of U is:

$$\hat{\sigma}_U = \sqrt{\frac{50^2(0.05) + 10^2(0.575) + 1.267}{5}} = 6.062.$$

TABLE 8-4. Demerits per Unit for Nonconformities in Automobiles

Sample number	Total demerits	Demerits per unit	Sample number	Total demerits	Demerits per unit
1	58	11.6	14	82	16.4
2	32	6.4	15	28	5.6
3	56	11.2	16	43	8.6
4	71	14.2	17	55	11.0
5	68	13.6	18	32	6.4
6	33	6.6	19	58	11.6
7	20	4.0	20	26	5.2
8	75	15.0	21	64	12.8
9	49	9.8	22	40	8.0
10	166	33.2	23	32	6.4
11	182	16.4	24	47	9.4
12	58	11.6	25	24	4.8
13	9	1.8			

Hence, the revised control limits are:

$$9.517 \pm 3(6.062) = 9.517 \pm 18.186 = (0, 27.703).$$

All of the points lie within the revised limits.

8-33. The centerline of a demerits per unit chart with the specified weights is:

$$\overline{U} = 10(0.064) + 5(0.6) + 1(1.264) = 4.904.$$

The standard deviation of U is:

$$\hat{\sigma}_U = \sqrt{\frac{10^2(0.064) + 5^2(0.6) + 1.264}{5}} = 2.129.$$

So, the control limits are:

$$4.904 \pm 3(2.129) = 4.904 \pm 6.387 = (0, 11.291).$$

The demerits per unit for each sample are shown in Table 8-5. All of the observations fall within the control limits.

TABLE 8-5. Demerits per Unit for Nonconformities in Automobiles

Sample number	Total demerits	Demerits per unit	Sample number	Total demerits	Demerits per unit
1	33	6.6	14	47	9.4
2	17	3.4	15	18	3.6
3	16	3.2	16	23	4.6
4	21	4.2	17	15	3.0
5	38	7.6	18	17	3.4
6	18	3.6	19	33	6.6
7	15	3.0	20	16	3.2
8	25	5.0	21	19	3.8
9	29	5.8	22	25	5.0
10	56	11.2	23	22	4.4
11	27	5.4	24	27	5.4
12	33	6.6	25	14	2.8
13	9	1.8			

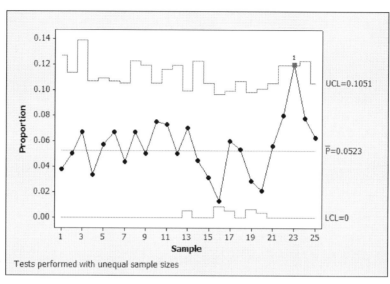

FIGURE 8-9. p-chart of significant medication errors

8-34. a)   A p-chart for the proportion of significant medication errors is constructed and shown in Figure 8-9. From the p-chart, the centerline = 0.0523, while the UCL and LCL vary by sample. Sample 23 plots above the UCL and the process is out of control.

b)   Deleting sample 23, the revised p-chart is constructed and shown in Figure 8-10. The revised centerline = 0.0503, while the UCL and LCL vary by sample. The process is now in control.

c)   From this stable process, the expected proportion of significant medication errors is 5.03%.

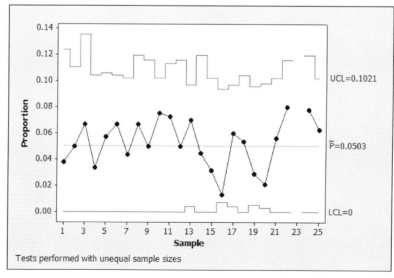

FIGURE 8-10. Revised p-chart of significant medication errors

144

d) Estimate of the approximate process standard deviation from the revised chart = (0.1021 - 0.0503)/3 = 0.0173. Based on a goal value of 1%, the current process average is about 2.33 standard deviations from the goal value.

$$Z = \frac{0.0503 - 0.01}{0.0173} = 2.33.$$

Hence, this goal would be difficult to achieve. Management needs to critically look at the process to identify safeguards that will detect medication errors. This may require major modifications to the current process.

8-35. The revised centerline was found as 0.0333, and the revised control limits were found to be (0, 0.0871). The probability of not detecting a shift, given a value of the process average p, which is the probability of a type II error, is found from:

$$\beta = P(X < 100(0.0871) \mid p) - P(X \le 100(0) \mid p)$$

$$= P(X \le 8 \mid p) - P(X \le 0 \mid p).$$

Since n is 100, we use the Poisson approximation to the Binomial distribution. Table 8-6 shows the value of β for different values of p. The OC curve is a plot of β versus p.

From the table, for p = 0.07, the probability of not detecting this shift on the first sample drawn after the change is 0.728. The probability of detecting the shift by the third sample is 0.272 + (0.728)(0.272) + (0.728)(0.728)(0.272) = 0.6142.

TABLE 8-6. Probability of a Type II Error for Various Values of p

p	$P(X \le 8 \mid p)$	$P(X \le 0 \mid p)$	P (type II error) = β
0.04	0.979	0.018	0.961
0.045	0.960	0.011	0.949
0.05	0.932	0.007	0.925
0.06	0.847	0.002	0.845
0.07	0.729	0.001	0.728
0.10	0.333	0.000	0.333
0.14	0.062	0.000	0.062
0.16	0.022	0.000	0.022

8-36. The revised centerline for a c-chart from Problem 8-27 is $\bar{c} = 6.875$ dietary errors per 100 trays, with the revised control limits being (0, 14.741). For a given value of the process average, the probability of a type II error is:

$$\beta = P[X < 14.741 \mid c] - P[X \le 0 \mid c]$$

$$= P[X \le 14 \mid c] - P[X \le 0 \mid c].$$

Using the Poisson tables, values of β are computed for different values of c and are shown in Table 8-7. If the process average increase to 10 errors per 100 trays, $\beta = 0.917$. So, the probability of detecting this change on the first sample drawn after the change $= (1 - 0.917) = 0.083$. The OC curve is a plot of β versus the various values of c.

8-37. The two-sigma control limits are:

$$6.875 \pm 2\sqrt{6.875} = 6.875 \pm 5.244 = (1.631, 12.119).$$

The probability of a type II error, when the process average changes, is:

$$\beta = P[X < 12.119 \mid c] - P[X \le 1.631 \mid c]$$
$$= P[X \le 12 \mid c] - P[X \le 1 \mid c].$$

For the process average of 8 dietary errors per 100 trays, we have:

$$\beta = P[X \le 12 \mid c = 8] - P[X \le 1 \mid c = 8]$$
$$= 0.936 - 0.003 = 0.933.$$

TABLE 8-7. **Probability of a Type II Error for Various Value of c**

c	$P[X \le 14 \mid c]$	$P[X \le 0 \mid c]$	$\beta = P$ (type II error)
7	0.994	0.001	0.993
8	0.983	0.000	0.983
9	0.959	0.000	0.959
10	0.917	0.000	0.917
12	0.772	0.000	0.772
14	0.570	0.000	0.570
16	0.368	0.000	0.368
18	0.208	0.000	0.208
20	0.105	0.000	0.105

Hence, the probability of detecting the change on the first sample drawn after the change = 1 − 0.933 = 0.067. We would prefer the two-sigma control limits when we desire to detect small changes in the process average as soon as possible, at the expense of an increased type I error.

8-38. The number of procedures until a complication occurs is monitored using a geometric distribution and the historical complication rate of 0.1%. The centerline and control limits for $\alpha = 0.005$ are:

$$\text{Centerline} = \text{CL} = \frac{1}{0.001} = 1000$$

$$\text{UCL} = \frac{\ln(0.0025)}{\ln(0.999)} = 5961.66$$

$$\text{LCL} = \frac{\ln(0.9975)}{\ln(0.999)} = 2.49.$$

If a plot is constructed of the observed number of procedures until a complication, with the vertical axis on a logarithmic scale, no points are outside the control limits and no discernible patterns are found. We conclude that the process is in control for an assumed complication rate of 0.1%.

If the type I error rate is 0.05, the control limits are:

$$\text{UCL} = \frac{\ln(0.025)}{\ln(0.999)} = 3687.04$$

$$\text{LCL} = \frac{\ln(0.975)}{\ln(0.999)} = 25.31.$$

The control limits move closer to the centerline. The probability of a false alarm will increase, if the complication rate stays at 0.1%. However, in the event the complicate rate changes from the value of 0.1%, the chances of detection increase. The selection of the level of the type I error could be influenced by the cost associated with a false alarm when the complicate rate is 0.1%. Further it could also be affected by the cost associated with the failure to detect a change in the complication rate from 0.1%, when a change takes place. In this situation, actions that need to be taken in the event of an increase in the complication rate, will not be initiated if the probability of a type II error increases.

8-39. For $\alpha = 0.005$, using a one-sided limit to detect an improvement, we have:

$$n > \frac{\ln(0.005)}{\ln(0.999)} = 5295.67 \simeq 5296.$$

For $\alpha = 0.05$, using a one-sided limit to detect an improvement, we have:

$$n > \frac{\ln(0.05)}{\ln(0.999)} = 2994.24 \simeq 2995.$$

For a larger tolerable level of false alarm, the minimum sample size decreases.

8-40. We need to determine the sample size, n, such that LCL > 0 for a p-chart. For two-sided limits, for $\alpha = 0.005$, $\alpha/2 = 0.0025$, yielding a Z value of -2.81. We have:

$$0.001 - 2.81 \sqrt{\frac{(.001)(.999)}{n}} > 0$$

or $n > \dfrac{(2.81)^2 (.001)(.999)}{(0.001)^2} = 7888.20 \simeq 7889.$

8-41. Using a complication rate of 0.2% and $\alpha = 0.005$, the centerline and control limits are:

$$\text{Centerline} = \text{CL} = \frac{1}{0.002} = 500$$

$$\text{UCL} = \frac{\ln(0.0025)}{\ln(0.998)} = 2992.74$$

$$\text{LCL} = \frac{\ln(0.9975)}{\ln(0.998)} = 1.25.$$

Using a complication rate of 0.5% and $\alpha = 0.005$, the centerline and control limits are:

$$\text{Centerline} = \text{CL} = \frac{1}{0.005} = 200$$

$$\text{UCL} = \frac{\ln(0.0025)}{\ln(0.995)} = 1195.29$$

$$\text{LCL} = \frac{\ln(0.9975)}{\ln(0.995)} = 0.50.$$

8-42. Using a complication rate of 0.1% and $\alpha = 0.005$, the UCL was found to be 5961.66. If the UCL is reduced by half, we use a UCL = 2980.83. We have:

$$2980.83 = \frac{\ln(0.0025)}{\ln(1-p)}$$

$\ln(1-p) = -0.00201$, yields $p = 1 - 0.997992 = 0.002008 = 0.2008\%$.

8-43. Using an exponential distribution, with parameter $\lambda = 0.001$ and $\alpha = 0.005$, for the time interval between complications, we have:

$$\text{Centerline} = CL = \frac{0.6931}{0.001} = 693.1$$

$$LCL = -\frac{1}{0.001}\ln(1-0.0025) = 2.50$$

$$UCL = -\frac{1}{0.001}\ln(0.0025) = 5991.46.$$

All of the observations are within the control limits.

8-44. a)   A u-chart is constructed for the number of items not checked per restrained patient. The centerline is CL = 0.0708, with the UCL and LCL varying by sample. Figure 8-11 shows the u-chart. Sample 13 plots above the UCL and so the process is out-of-control.

b)   Deleting sample 13, a revised u-chart is constructed and shown in Figure 8-12.

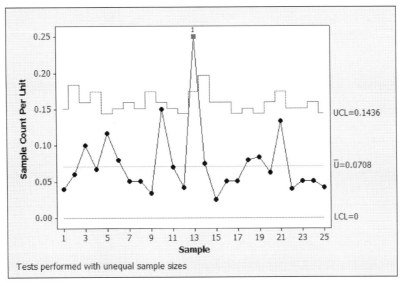

FIGURE 8-11. u-chart of number of items not checked

149

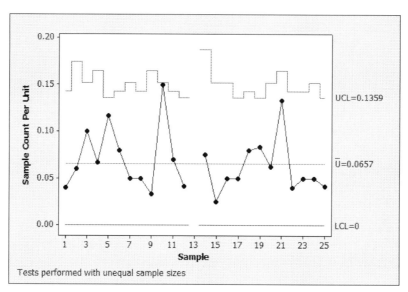

FIGURE 8-12. Revised u-chart of number of items not checked

The revised centerline is CL = 0.0657, with the UCL and LCL varying by sample.

c) Using the centerline of the revised u-chart, the expected number of items not checked per restrained patient = 0.0657.

CHAPTER 9

PROCESS CAPABILITY ANALYSIS

9-1.   Specification limits are determined by the needs of the customer. These are bounds placed on product or service characteristics to ensure adequate functioning of the product or meeting the service expectations of the consumer. Control limits, on the other hand, represent the variation between samples or subgroups, of the statistic being monitored (such as sample average). Control limits have no relationship to the specification limits.

9-2.   Natural tolerance limits represent the inherent variation, when special causes are removed, in the quality characteristic of individual product/service items. Natural tolerance limits are usually found based on the assumption of normality of distribution of the characteristic. Ideally, they should be found based on the actual distribution of the characteristic. Specification limits have been defined in Exercise 9-1. A process capability index incorporates both the specification limits and the natural tolerance limits. It determines the ability of the process to meet the specification limits, thus indicating a measure of goodness of the process.

9-3.   Statistical tolerance limits define bounds of an interval that contains a specified proportion $(1-\alpha)$ of the population with a given level of confidence (γ). These bounds are found using sample statistics, for example, the sample mean and sample standard deviation. As the sample size becomes large, the statistical tolerance limits approach the values that are found using the population parameters (population mean and population standard deviation, for example). Statistical tolerance limits are usually found based on a normal distribution or using nonparametric methods. Natural tolerance limits have been defined in Exercise 9-2. These limits, for an in-control process, represent coverage such that just about all, or 99.74% using the normality assumption, of the distribution is contained within these bounds.

9-4.   It is possible for a process to be in control, when only common causes prevail, and still produce nonconforming output that does not meet the specification limits. This implies that the inherent variation of the quality characteristic in the process, when it is in control, as determined by the spread between the natural tolerance limits, exceeds the spread between the specification limits. Some corrective measures could be to explore if the customer is willing to loosen the specification limits, reduce the process spread through better equipment, better raw material, or better personnel, or in the short run to shift the process average so as to reduce the total cost of nonconformance (which could be the cost of rework and scrap in a product situation).

9-5.   When the process spread is less than the specification spread, one main advantage is that if the process mean does not change and is centered between the specification limits, just about all of the items will be acceptable to the customer. This makes an assumption of normality of distribution of the quality characteristic. C_p should be greater than 1. It is possible for C_{pk} to be ≤ 1, if the process mean is closer to one of the specification limits by less than 3σ.

9-6.   The index C_{pk} measures actual process performance (proportion nonconforming) and is influenced by both the process mean and the process standard deviation. The index C_{pm}

incorporates both the process mean and the process standard deviation and is a measure that represents the deviation of the process mean from the target value. The index C_{pmk} incorporates process mean and process variability as well as the deviation of the process mean from the target value. When the process mean is at the target value, $C_{pmk} = C_{pk}$. It is known that $C_p \geq C_{pk} \geq C_{pmk}$ and $C_p \geq C_{pm} \geq C_{pmk}$. The index C_{pq} may be used when the distribution of the quality characteristic is not normal. When only an empirical distribution of the characteristic is available based on the observations, C_{pq} uses a nonparametric approach and estimates percentiles based on the observed empirical distribution. This approach may therefore be used for any distribution of the characteristic.

9-7.   The process must be in control prior to estimating its capability. Depending on the type of control chart used, estimates of the process mean and standard deviation and thereby capability can be developed. If $\bar{X}$ and R-charts are used, estimates are:

$$\hat{\mu} = \bar{\bar{X}}, \quad \hat{\sigma} = \frac{\bar{R}}{d_2}.$$

If $\bar{X}$ and s charts are used, estimates are:

$$\hat{\mu} = \bar{\bar{X}}, \quad \hat{\sigma} = \frac{\bar{s}}{c_4}.$$

If X and MR-charts are used, estimates are:

$$\hat{\mu} = \bar{X}, \quad \hat{\sigma} = \frac{\overline{MR}}{d_2}.$$

For attribute charts, the centerline is a measure of the capability of the process. So, for a p-chart, the centerline $\bar{p}$, for a c-chart, the centerline $\bar{c}$, for a u-chart, the centerline $\bar{u}$, and for a U-chart, the centerline $\bar{U}$ are corresponding measures.

9-8.   Let Y represent the project completion time, with X_i representing the ith individual operation time. We have:

$$Y = X_1 + X_2 + ... + X_k,$$

where k represents the number of individual operations. The mean project completion time is:

$$\mu_Y = \mu_1 + \mu_2 + ... + \mu_k,$$

where μ_i represents the mean of the ith operation.

If the operations are independent, the variance of the project completion time is:

$$\text{Var }(Y) = \sigma_1^2 + \sigma_2^2 + ... + \sigma_k^2,$$

where σ_i^2 represents the variance of the ith operation.

In the event that the operations are not independent, the variance of the project completion time is given by:

$$\text{Var }(Y) = \sigma_1^2 + \sigma_2^2 + ... + \sigma_k^2 + 2\sum_{i=1}^{k-1}\sum_{j=i+1}^{k} \text{Cov }(X_i, X_j),$$

where $\text{Cov }(X_i, X_j)$ represents the covariance between operation i and j completion times. The mean project completion time will be the same as before.

9-9. Let Y represent the assembly dimension with X_1 and X_2 representing the two component dimensions. We have:

$$Y = X_1 - X_2.$$

Let the component means be given by μ_1 and μ_2 and the component variances be denoted by σ^2, being equal for both components. So:

$$\sigma_Y^2 = Var (Y) = \sigma^2 + \sigma^2 = 2\sigma^2$$

or $\sigma^2 = \sigma_Y^2 / 2$ $\quad or \quad \sigma = \sigma_Y / \sqrt{2}.$

Assuming tolerances on Y are given, if the process is just barely capable:

$6\sigma_Y = \text{USL} - \text{LSL} = \text{T, or } \sigma_Y = \text{T/6.}$

Hence:

$$\sigma = \sigma_Y / \sqrt{2} = T / (6\sqrt{2}).$$

Now, the tolerances for each component, assuming normality, are:

For X_1: $\mu_1 \pm 3\sigma = \mu_1 \pm 3T / (6\sqrt{2}) = \mu_1 \pm T / (2\sqrt{2}).$

For X_2: $\mu_2 \pm 3\sigma = \mu_2 \pm 3T/(6\sqrt{2}) = \mu_2 \pm T/(2\sqrt{2})$.

9-10. The traditional process capability measures, for example $C_p, C_{pk}, C_{pm}, C_{pmk}$ assume normality of distribution of the quality characteristic. If the distribution of the characteristic deviates drastically from the normal distribution, the above measures may not be appropriate to make inferences. In particular, it is important to identify the distribution of the characteristic to determine the degree of nonconformance. In the context of waiting time for service, in a fast-food restaurant, this distribution could be exponential. One measure could be the C_{pq} index, given the specifications on waiting time, where the percentiles would be obtained from the empirical distribution of waiting time. Alternatively, one could calculate:

P [Waiting time > USL], given an estimate of the parameter of the distribution from the observed data.

A refined model could incorporate the distribution of the arrival of customers during the lunch-hour, and the distribution of service times during the lunch-hour to determine the distribution of waiting time during the lunch-hour.

9-11. All measuring instruments, such as those to measure unloading times of supertankers, have an inherent variability, known as repeatability. Further, variation between operators who unload and the interaction between operators and the type of cargo that is unloaded is known as reproducibility. In addition to the process variation, repeatability and reproducibility, will also have an influence on the observed capability index (C_p^*). It is known that:

$$C_p^* = \frac{USL - LSL}{6\sigma_m} = \frac{1}{\sqrt{(1/C_p)^2 + r^2}},$$

where C_p is the true capability index and

$$r = \frac{6\sigma_e}{USL - LSL},$$

where σ_e represents the standard deviation of the measurement error due to repeatability and reproducibility.

When $r = 0$, $C_p^* = C_p$. However, this is not realistic since some degree of measurement error always exists. Usually, $C_p > C_p^*$. Also, if the process variability approaches zero, an upper bound on C_p^* is:

$$C_p^* \leq \frac{1}{r},$$

where r is the precision-to-tolerance ratio.

9-12. This has been discussed in Exercise 9-11. Gage repeatability measures the variation in unloading times of supertankers, when using the same operator, same measuring instrument, and the same set of tasks in unloading. Reproducibility measures the variation between operators, when unloading the same shipment using the same measuring instrument. It also measures the interaction that may exist between operators and the type of cargo that is being unloaded.

9-13. The natural tolerance limits are $44 \pm 3(3) = (35, 53)$. The standard normal values at the lower and upper specification limits are:

$$Z_1 = (40 - 44)/3 = -1.33$$

$$Z_2 = (55 - 44)/3 = 3.67.$$

Using standard normal tables, the proportion below the lower specification limits is 0.0918, while the proportion above the upper specification limit is 0.0000.

9-14. The C_p index is computed as:

$$C_p = (55 - 40)/[6/(3)] = 0.833.$$

Since C_p is less than 1, we have an undesirable situation. All of the output from the process will not meet specifications. The capability ratio is:

$$CR = 6(3)/(55 - 40) = 1.2.$$

So, the process uses up 120% of the specification range.

The process mean should be shifted to 47.5, which is the midpoint between the specification limits. The standard normal values at each specification limit will be ± 2.5. The proportion outside each specification limit will be 0.0062, making the total proportion of nonconforming product as 0.0124.

9-15. Upper capability index is:

$$CPU = (3.5 - 2.3)/(3(0.5)) = 0.8.$$

Since CPU is less than 1, the emergency service unit will not fully meet the desirable goal. To determine the proportion of patients who will have to wait longer than the specified goal, assuming a normal distribution of waiting time, the standard normal value is:

$$Z = (3.5 - 2.3)/0.5 = 2.4.$$

The proportion of patients who will have to wait longer than 3.5 minutes is 0.0082. Some remedial actions may include increasing the number of support staff in charge of admissions and preparation for treatment, and expanding the size of the emergency unit if data supports increased usage.

9-16.　$C_{pk} = \min \left[\dfrac{55 - 44}{3(3)}, \dfrac{44 - 40}{3(3)} \right] = \min \{1.222, 0.444\} = 0.444.$

Since C_{pk} is less than 1, an undesirable situation exists. Since the target value is 47.5, $\delta = (\mu - \tau)/\sigma = (44 - 47.5)/3 = -1.167.$

$C_{pm} = C_p / \sqrt{1 + \delta^2} = 0.833 / \sqrt{1 + (-1.167)^2} = 0.542.$

Note that the process mean is off from the desirable target value by 1.167 standard deviations.

$C_{pmk} = \dfrac{\min[(55 - 44), (44 - 40)]}{3\sqrt{3^2 + (44 - 47.5)^2}}$

$= 0.289.$

The value of C_{pmk} incorporates not only the process mean and variability and the location of the process mean relative to the closest specification limit, it also takes into account the deviation of the process mean from the target value.

If the process mean shifts to the midpoint between the specification limits, it will be at 47.5. The standard normal value would be ± 2.5 at the specification limits, making the total proportion nonconforming as $2(0.0062) = 0.0124$. In Exercise 9-13, the current proportion nonconforming was found as 0.0918. So, an improvement has occurred.

9-17.　a)　　$C_{pk} = \min \left[\dfrac{125 - 122}{3(2)}, \dfrac{122 - 115}{3(2)} \right] = \min \{0.5, 1.167\} = 0.5.$

Since C_{pk} is less than 1, an undesirable situation exists. The standard normal values at the specification limits are:

$Z_1 = (115 - 122)/2 = -3.5$

$Z_2 = (125 - 122)/2 = 1.5.$

The proportion below the LSL is 0.0000, while the proportion above the USL is 0.0668.

Since the target value is 120, $\tau^2 = 2^2 + (122 - 120)^2 = 8$. So $\tau = \sqrt{8} = 2.8284$.

$C_{pm} = (125 - 115)/[6(2.8284)] = 0.589$.

Note that the process mean is off from the target value by 1 standard deviation.

$$C_{pmk} = \frac{\min[(125 - 122), (122 - 115)]}{3\sqrt{2^2 + (122 - 120)^2}}$$

$= 0.353$.

The small value of C_{pmk} is an indication of the process mean being close to one of the specification limits, in standardized units, and also deviating from the target value.

b) If the process mean is set at the target value of 120, the standard normal value would be ± 2.5 at the specification limits. The total proportion nonconforming would be $2(0.0062) = 0.0124$. So, the reduction in the fraction nonconforming would be $= 0.0668 - 0.0124 = 0.0544$.

At the current setting of the process mean at 122, the daily costs are as follows:

Cost of parts below LSL: 0.0000 x $30,000$ x $1.00 = \$0.00$
Cost of parts above USL: 0.0668 x $30,000$ x $0.50 = \$1002.00$

That daily total cost of nonconformance is $1002.00

With the process mean at 120, the daily costs are as follows:

Cost of parts below LSL: 0.0062 x $30,000$ x $1.00 = \$186.00$
Cost of parts above USL: 0.0062 x $30,000$ x $0.50 = \$93.00$

The total daily cost of nonconformance is $279.

9-18. a) A normal probability plot is shown in Figure 9-1. The p-value using the Anderson-Darling test is $0.129 > \alpha = 0.05$. So, we do not reject the null hypothesis of normality.

b) The sample mean waiting time is found to be 2.906 minutes with a standard deviation of 1.327 minutes.

c) CPU = (4 – 2.906)/(3(1.327)) = 0.275. Since CPU < 1, an undesirable situation exists. With no lower specification limit, C_{pk} will have the same value as CPU. The standard normal value at USL is Z = (4 – 2.906)/1.327 = 0.824 $\simeq$ 0.82, with the proportion above USL being 0.2061. Hence, 20.61% of the customers will have to wait more than 4 minutes.

9-19. a) The centerline on the $\overline{X}$-chart is 1000/25 = 40, while the centerline on the R-chart is 250/25 = 10. Control limits for an $\overline{X}$-chart are: 40 $\pm$ 0.729(10) = (32.71, 47.29). The lower control limit for an R-chart is 0, while the upper control limit is 2.282(10) = 22.82.

b) An estimate of the process standard deviation is $\hat{\sigma}$ = 10/2.059 = 4.857. The upper capability index is:

$$CPU = \frac{50 - 40}{3(4.857)} = 0.686.$$

Since CPU is less than 1, an undesirable situation exists.

c) The standard normal value at USL is Z = (50 – 40)/4.857 = 2.058 $\simeq$ 2.06. The proportion of customers who will have to wait more than 50 minutes is 0.0197.

d) If the mean waiting time is reduced to 35 minutes, the standard normal value at USL is Z = (50 – 35)/4.857 = 3.088 $\simeq$ 3.09. The proportion of customers who will still have to wait more than 50 minutes is 0.0010.

9-20. a) The centerline on an $\overline{X}$-chart is 2550/30 = 85, while that on an s-chart is 195/30

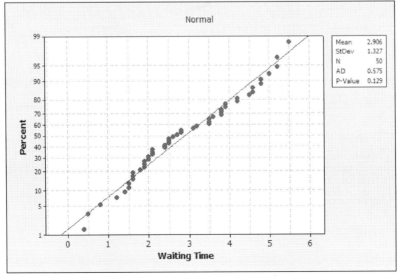

FIGURE 9-1. Normal probability plot of waiting time

= 6.5. The control limits for an s-chart are: LCL = 0(6.5) = 0; UCL = 2.089(6.5) = 13.578. The control limits for an $\overline{X}$-chart are: LCL = 85 − 1.427(6.5) = 75.7245; UCL = 85 + 1.427(6.5) = 94.2755.

b)   The process mean is estimated as 85. The process standard deviation is estimated as 6.5/0.9400 = 6.915.

c) $C_p = \dfrac{105 - 75}{6(6.915)} = 0.723.$

$$C_{pk} = \min \left[\frac{105 - 85}{3(6.915)}, \; \frac{85 - 75}{3(6.915)} \right] = 0.482.$$

Since C_p and C_{pk} are less than 1, an undesirable situation exists.

d)   Since the target value is 90, $\delta = (85 - 90)/6.915 = -0.723.$

$$C_{pm} = C_p / \sqrt{1 + \delta^2} = 0.723 / \sqrt{1 + 0.5228} = 0.586.$$

$$C_{pmk} = \frac{\min[(105 - 85), (85 - 75)]}{3\sqrt{6.915^2 + (85 - 90)^2}}$$

$$= 0.391.$$

e)   The standard normal values at the specification limits are:

$Z_1 = (75 - 85)/6.915 = -1.446 \simeq -1.45$

$Z_2 = (105 - 85)/6.915 = 2.892 \simeq 2.89.$

The proportion of the product below the LSL is 0.0735, while the proportion above the USL is 0.0019, making the proportion nonconforming as 0.0754.

f)   If the process mean is moved to 88, the standard normal values at the specification limits are:

$Z_1 = (75 - 88)/6.915 = -1.88$

$Z_2 = (105 - 88)/6.915 = 2.46.$

The proportion of the product below the LSL now is 0.0301, while the proportion above the USL is 0.0069, making the proportion nonconforming as 0.0370. One proposal is to shift the process mean toward the target value of 90. Another is to determine process parameter settings that will reduce process variability.

9-21. a) Control charts for the mean, $\overline{X}$, and range, R, are constructed. The centerline on the R-chart is:

$$\overline{R} = \frac{144}{25} = 5.760.$$

The upper and lower control limits on the R-chart are:

$$UCL = D_4\overline{R} = (2.114)(5.760) = 12.177$$

$$LCL = D_3\overline{R} = 0(5.760) = 0.0.$$

Next, the centerline on the $\overline{X}$-chart is calculated:

$$\overline{\overline{X}} = \frac{582}{25} = 23.280.$$

The upper and lower control limits on the $\overline{X}$-chart are:

$$UCL = \overline{\overline{X}} + A_2\overline{R} = 23.280 + (0.577)(5.76) = 26.603$$

$$LCL = \overline{\overline{X}} - A_2\overline{R} = 23.280 - (0.577)(5.76) = 19.956.$$

Observation 7 plots above the mean on the $\overline{X}$-chart, indicating an out-of-control condition.

b) Assuming that special causes have been identified and remedial actions taken, the revised centerline on the $\overline{X}$ and $\overline{R}$-charts are:

$$\overline{\overline{X}} = \frac{582 - 29}{24} = 23.042$$

$$\overline{R} = \frac{144 - 7}{24} = 5.708.$$

The revised limits on the $\overline{X}$-chart are:

$$23.042 \pm (0.577)(5.708) = (19.749, 26.335).$$

The revised limits on the R-chart are:

$$\text{UCL} = D_4 \bar{R} = (2.114)(5.708) = 12.067$$

$$\text{LCL} = D_3 \bar{R} = 0(5.708) = 0.$$

Estimate of the process mean = 23.042.

Estimate of the process standard deviation is:

$$\hat{\sigma} = \frac{5.708}{2.326} = 2.454.$$

c)   The process capability indices are calculated:

$$C_p = \frac{29 - 17}{6(2.454)} = 0.815.$$

$$C_{pk} = \frac{\min[(29 - 23.042), (23.042 - 17)]}{3(2.454)}$$

$$= 0.809.$$

For a target value of 23, $\delta = (23.042 - 23)/2.454 = 0.0171$.

$$C_{pm} = \frac{0.815}{\sqrt{1 + (0.0171)^2}} = 0.8149.$$

$$C_{pmk} = \frac{0.809}{\sqrt{1 + (0.0171)^2}}$$

$$= 0.8089.$$

d)   The standard normal values at the specification limits are:

$$Z_1 = \frac{17 - 23.042}{2.454} = -2.462 \simeq 2.46$$

$$Z_2 = \frac{29 - 23.042}{2.454} = 2.428 \simeq 2.43.$$

From the standard normal tables, the proportion below the lower specification limit is 0.0069, while the proportion above the upper specification limit is 0.0075. The total proportion that does not meet specifications is 0.0144, implying that the proportion that meets government standards is 0.9856.

e)   A 95% confidence interval for C_{pk}, under the assumption of normality is:

$$0.809 \pm 1.96 \sqrt{\frac{1}{9(25)} + \frac{0.809^2}{2(24)}}$$

$$0.809 \pm 0.2635 = (0.5455, 1.0725).$$

f) $H_o : C_{pk} \geq 1.00$ vs. $H_a : C_{pk} < 1.00$. Using $\alpha = 0.05$, we calculate a one-sided UCL:

$$UCL = 0.809 + 1.645 \sqrt{\frac{1}{9(25)} + \frac{0.809^2}{2(24)}}$$

$$= 0.809 + 0.2212 = 1.0319.$$

Since the hypothesized value of $C_{pk} = 1.00 < UCL = 1.0319$, we do not reject H_o. Hence, we cannot conclude that $C_{pk} < 1$.

9-22.   The standard deviations of the components are found below:

Component A: $\sigma_1 = (10.5 - 9.5)/6 = 0.167$
Component B: $\sigma_2 = (4.2 - 3.8)/6 = 0.067$
Component C; $\sigma_3 = (5.1 - 4.9)/6 = 0.033$.

The mean gap length $= 10 - 4 - 5 = 1$ cm.

The variance of the gap length $= (0.167)^2 + (0.067)^2 + (0.033)^2 = 0.03347$, yielding a standard deviation of 0.183. Now, the tolerances of the gap are: $1 \pm 3(0.183) = 1 \pm 0.549 = (0.451, 1.549)$.

The standard normal values at the specification limits are:

$Z_1 = (0.699 - 1)/0.183 = -1.64$

$Z_2 = (1.101 - 1)/0.183 = 0.55$.

The proportion of assemblies below the LSL is 0.0505, while that above the USL is 0.2912, making the fraction nonconforming as 0.3417. One way to reduce the fraction nonconforming is to increase the mean length of B or C, so that the mean gap length is reduced to the target value of 0.9. Alternatively, measures to reduce the variability of the operations should be investigated.

9-23. The standard normal values at the specification limits are:

$Z_1 = (0.9 - 1)/0.183 = -0.55$

$Z_2 = (1.20 - 1)/0.183 = 1.09.$

The proportion below the LSL is 0.2912, while that above the USL is 0.1379.

Daily cost of rework = 2000 x 0.15 x 0.2912 = $87.36
Daily cost of scrap = 2000 x 0.40 x 0.1379 = $110.32.

The daily total cost of rework and scrap is $197.68. One way to reduce the cost is to reduce the mean dimensions of B and C to bring the mean value of gap to the target value of 1.05. Alternatively, measures to reduce variability of the operations should be investigated.

9-24. The specifications are $35 \pm 0.5 = (34.5, 35.5)$. The standard deviation of the length of the assembly is estimated as: $(35.5 - 34.5)/6 = 0.167$. Under the assumption of equal variances of each of the components, we have $4\sigma_1^2 = (0.167)^2$ or $\sigma_1^2 = 0.00694$, yielding a standard deviation of each component as 0.083. So, the tolerances on the components are:

A: $3 \pm 3(0.083) = (2.751, 3.249)$
B: $8 \pm 3(0.083) = (7.751, 8.249)$
C: $10 \pm 3(0.083) = (9.751, 10.249)$
D: $14 \pm 3(0.083) = (13.751, 14.249).$

9-25. The standard deviation of the length of the assembly is estimated as $(35.3 - 34.7)/6 = 0.1$. Let σ_1 denote the standard deviation of component A, which is the same as that of component C. The standard deviation of component B is $2\sigma_1$, which is also the same as that of component D. Hence, $\sigma_1^2 + 4\sigma_1^2 + \sigma_1^2 + 4\sigma_1^2 = (0.1)^2$, yielding $\sigma_1 = 0.0316$. The tolerances for each component are as follows:

A: $3 \pm 3(0.0316) = (2.9052, 3.0948)$
B: $8 \pm 3(0.0632) = (7.8104, 8.1896)$
C: $10 \pm 3(0.0316) = (9.9052, 10.0948)$
D: $14 \pm 3(0.0632) = (13.8104, 14.1896).$

9-26. The standard deviation of $X_1 = 0.15/3 = 0.05$, while the standard deviation of $X_2 = 0.05/3 = 0.0167$. The mean dimension of Y = 12 − 5 = 7 cm. The variance of Y = $(0.05)^2 + (0.0167)^2 = 0.002779$, yielding a standard deviation of Y = 0.0527 cm. So, the

164

specifications for Y are: $7 \pm 3(0.0527) = (6.8419, 7.1581)$. The standard normal value at 7.10 is $Z = (7.10 - 7)/0.0527 = 1.90$, yielding a proportion above the specified limit to be 0.0287.

9-27. The standard deviation of $Y = 0.2/3 = 0.0667$. Denoting σ_2^2 as the variance of X_2 , we have:

$$3\sigma_2^2 + \sigma_2^2 = (0.0667)^2, \text{ yielding } \sigma_2 = 0.0333.$$

So, the standard deviation of $X_1 = \sqrt{3} \ (0.0333) = 0.0578$.

Hence, tolerances for the components are:

X_1 : $14 \pm 3(0.0578) = (13.8266, 14.1734)$

X_2 : $8 \pm 3(0.0333) = (7.9001, 8.0999)$.

9-28. The standard deviation of each plate thickness $= 0.2/3 = 0.0667$. The mean thickness of the assembly $= 4(3) = 12$ cm. The variance of the assembly thickness $= 4(0.0667)^2 = 0.0178$, yielding a standard deviation of 0.1333 cm. Hence, tolerances for the thickness of the assembly are: $12 \pm 3(0.1333) = (11.6001, 12.3999)$.

9-29. a) The mean of the difference between the hole diameter and the shaft diameter $= 6.2 - 6 = 0.2$ cm. The standard deviation of the hole diameter $= 0.03/3 = 0.01$, while the standard deviation of the shaft diameter $= 0.06/3 = 0.02$. The variance of the difference between the hole and shaft diameter $= (0.01)^2 + (0.02)^2 = 0.0005$, yielding a standard deviation of 0.0224. To find the probability of a clearance fit, we must find the probability of the difference between the hole and shaft diameter being > 0 . The standard normal value at 0 is $Z = (0 - 0.2)/0.0224 = -8.92$. So, just about all the assemblies will have a clearance fit.

b) The probability of an assembly having an interference fit is just about 0.

9-30. Assuming that the specified tolerances are proportional to the size of the nominal dimension, the tolerances on the shaft radius are 2.5 ± 0.015 , and those for the hole radius are 2.625 ± 0.04 . Now, the mean clearance $= 2.625 - 2.5 = 0.125$ cm. The variance of the clearance $= (0.015/3)^2 + (0.04/3)^2 = 0.000203$, yielding a standard deviation of 0.0142. The standard normal values at the lower and upper specification limits are:

$Z_1 = (0.13 - 0.125)/0.0142 = 0.35$

$Z_2 = (0.23 - 0.125)/0.0142 = 7.39.$

The proportion of assemblies that will be acceptable = $(1 - 0.6368) = 0.3632$.

9-31. In Problem 9-30 the mean clearance was found to be 0.125 cm with a standard deviation of 0.0142. The standard normal value at 0.05 is $Z = (0.05 - 0.125)/0.0142 = -5.28$. So, the probability of a wobble is about 1.

9-32. Assume that the specified tolerances are proportional to the size of the nominal dimension. The mean of the difference between the cylinder and the piston radius is 0.05 cm. The standard deviation of the cylinder radius = $0.2/3 = 0.067$, while the standard deviation of the piston radius = $0.25/3 = 0.083$. The variance of the difference between the cylinder and piston radius = $(0.067)^2 + (0.083)^2 = 0.01138$, yielding a standard deviation of 0.1067. The standard normal value at 0 is $Z = (0 - 0.05)/0.1067 = -0.47$. Using the normal tables, the proportion of nonconforming assemblies is 0.3192. The standard normal value at 0.8 is $Z = (0.8 - 0.05)/0.1067 = 7.03$. The proportion of assemblies not meeting the stipulation is negligible.

9-33. a) The standard deviation of the operation times are computed as follows:

Operation 1: $\sigma_1 = 0.6/3 = 0.2$

Operation 2: $\sigma_2 = 0.6/3 = 0.2$

Operation 3: $\sigma_3 = 0.8/3 = 0.267$

Operation 4: $\sigma_4 = 0.3/3 = 0.1$.

The variance of the order completion time is:

$$\sigma^2 = (0.2)^2 + (0.2)^2 + (0.267)^2 + (0.1)^2$$

$$= 0.16129.$$

Hence, the standard deviation of the order completion time is:

$$\sigma = \sqrt{0.16129} = 0.4016.$$

Since the mean order completion time = 23 hours, the natural tolerance limits are:

$$23 \pm 3(0.4016) = 23 \pm 1.2048 = (21.7952, 24.2048).$$

b) The standard normal value at the goal is:

$$Z = \frac{23.5 - 23}{0.4016} = 1.245.$$

The proportion of orders that will take more than 23.5 hours is 0.1065.

c)   An upper capability index, CPU, would be appropriate here.

$$CPU = \frac{23.5 - 23}{3(0.4016)} = 0.415.$$

Since CPU is quite less than 1, quite a few orders (10.65%) will take more than the goal value.

d)   Assuming that the variance of the operation times are the same as before, the mean order completion now = 22 hours. The standard normal value at the goal is:

$$Z = \frac{23.5 - 22}{0.4016} = 3.735 \approx 3.74.$$

The proportion of orders that will take more than 23.5 hours is 0.0000, which is negligible.

9-34. Minitab software is used for the computations. Table 9-1 shows some of the gage repeatability and reproducibility statistics.

TABLE 9-1. Gage R&R indices

```
                              %Contribution
Source              VarComp   (of VarComp)
Total Gage R&R      0.0022420          4.59
  Repeatability     0.0021260          4.35
  Reproducibility   0.0001160          0.24
    Operator        0.0001160          0.24
Part-To-Part        0.0466183         95.41
Total Variation     0.0488603        100.00


Process tolerance = 0.1


                                 Study Var   %Study Var   %Tolerance
Source              StdDev (SD)  (6 * SD)        (%SV)    (SV/Toler)
Total Gage R&R       0.047350    0.28410         21.42       284.10
  Repeatability      0.046108    0.27665         20.86       276.65
  Reproducibility    0.010771    0.06463          4.87        64.63
    Operator         0.010771    0.06463          4.87        64.63
Part-To-Part         0.215913    1.29548         97.68      1295.48
Total Variation      0.221044    1.32626        100.00      1326.26


Number of Distinct Categories = 6
```

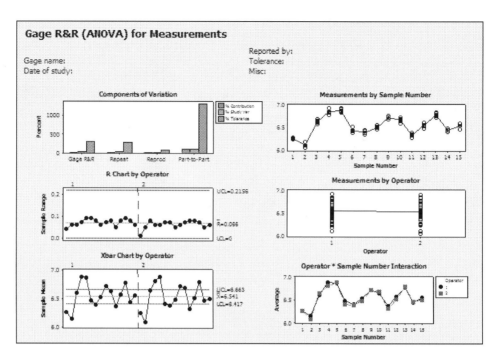

FIGURE 9-2. Gage repeatability and reproducibility for pH values

Figure 9-2 shows the results graphically for the gage R&R analysis. The operator by part interaction is not found to be significant (p-value = 0.718). Hence, the significant variance components are:

Repeatability ($\hat{\sigma}_t^2 = 0.002126$)

Reproducibility ($\hat{\sigma}_p^2 = 0.000116$) – variation due to operators

Total gage R&R ($\hat{\sigma}_e^2 = 0.002242$)

Process ($\hat{\sigma}^2 = 0.0466183$) – part-to-part variation

Total measured observations ($\hat{\sigma}_m^2 = 0.0488603$).

% Gage R&R = ($\hat{\sigma}_e / \hat{\sigma}_m$)100 = 21.42% (Table 9-1, under % Study Var). Since this is < 30%, the gage system is acceptable.

Precision to tolerance ratio (r) = $6\hat{\sigma}_e / (USL - LSL)$ = 2.841 = 284.10%. Since this ratio is not < 10%, the gage system is not capable of meeting the specified tolerances. Hence, a new gage system is necessary in order to meet the given tolerances.

% Process variation = ($\sigma_e / \hat{\sigma}$) 100

$$= (0.04735/0.21593) 100 = 21.93\%.$$

The gage system variability is about 22% of the process variability.

Number of distinct categories $= (\hat{\sigma} / \hat{\sigma}_e) \, 1.41$

$$= (0.215913/0.04735)(1.41) = 6.43 \simeq 6.$$

Since the number of distinct categories is $\geq$, 4, the gage system can adequately discern the observed values.

Hence, if the specified tolerance is a major requirement, the current gage system will not qualify.

The observed process potential is:

$$C_p{}^* = \frac{6.55 - 6.45}{6(0.221044)} = 0.075.$$

Given the very small value of $C_p{}^*$ relative to 1, the current gage system is not capable at all of meeting the specifications.

The true process potential, after discounting for the variability in the measurement system, is estimated as:

$$C_p = \frac{1}{\sqrt{(1/0.075)^2 - 2.841^2}} = 0.0768.$$

A 95% confidence interval for C_p is obtained as:

$$LCL = (0.075) \sqrt{\frac{\chi^2_{0.975,59}}{59}} = 0.075 \sqrt{\frac{39.6619}{59}} = 0.0615$$

$$UCL = (0.075) \sqrt{\frac{\chi^2_{0.025,59}}{59}} = 0.075 \sqrt{\frac{82.1174}{59}} = 0.0885.$$

9-35. a)   A normal probability plot is constructed and shown in Figure 9-3. Using the Anderson-Darling test for normality, the p-value is $0.046 < \alpha = 0.05$. Hence, we reject the null hypothesis of normality. Thus, capability analysis using the normal distribution would not be appropriate for the original data.

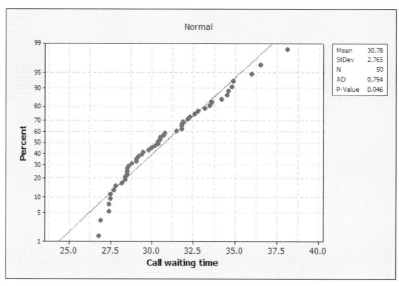

FIGURE 9-3. Normal probability plot of call waiting time

b) A Box-Cox transformation is considered and capability analysis is conducted on the transformed data and shown in Figure 9-4. Minitab identifies the optimal power coefficient (lambda) to be -2.1. The transformed data looks symmetric and close to a normal distribution. A test for normality for the transformed data passes the test. From the capability analysis, CPU = 0.48. The expected proportion above the USL is 0.0752.

c) Capability analysis is conducted using the Weibull distribution and shown in Figure 9-5. The long-term CPU index is 0.56, and the expected proportion above the USL is 0.0734.

d) A capability analysis using the normal distribution is conducted, even though it is

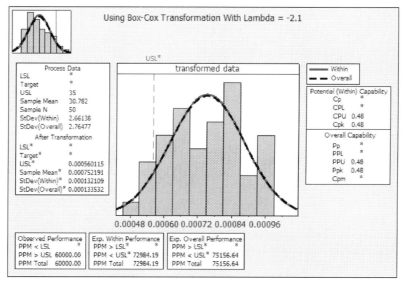

FIGURE 9-4. Capability analysis using Box-Cox transformation

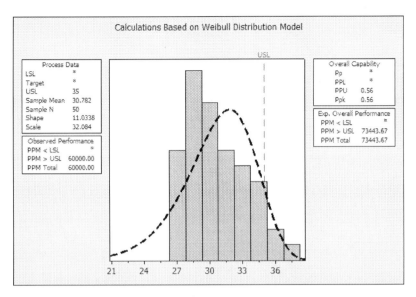

FIGURE 9-5. Capability analysis using Weibull model

not appropriate, to see the results. Figure 9-6 shows such an analysis. Note that the long-term CPU index is 0.53, while the expected proportion above the USL is 0.0636. These values should not be accepted and used since it has previously been determined that the test for normality of distribution failed. Note that the long-term CPU index is a bit inflated, compared to that obtained using the Box-Cox transformation.

9-36. Assuming that tolerances vary linearly with the dimension, the tolerances on the radius (r) are: 1 ± 0.03 cm, while the tolerances on height (h) are: 6 ± 0.06 cm. We have: $\mu_r = 1$, $\sigma_r = 0.03/3 = 0.01$, $\mu_h = 6$, $\sigma_h = 0.06/3 = 0.02$. The relationship for the volume of the cylindrical piece is:

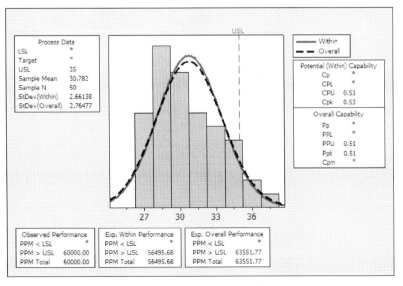

FIGURE 9-6. Capability analysis of call waiting time using normal distribution

171

$$V = \pi r^2 h$$

or mean volume $= \mu_v = \pi \mu_r^2 \ \mu_h$

$$= 3.1416(1)^2(6) = 18.8496 \ \text{cm}^3.$$

$$\frac{\partial v}{\partial r} = 2\pi rh, \frac{\partial v}{\partial h} = \pi r^2.$$

Thus, Variance (V) is given by:

$$\sigma_v^2 = (2\pi \ \mu_r \ \mu_h)^2 \ \sigma_r^2 + (\pi \ \mu_r^2)^2 \ \sigma_h^2$$

$$= (2\pi 6)^2 (0.01)^2 + \pi^2 (0.02)^2$$

$$= 0.1421 + 0.0039 = 0.146$$

$$\sigma_v = \sqrt{0.146} = 0.3821.$$

The natural tolerance limits on the volume of the cylinder, assuming normality, are:

$$18.8496 \pm 3(0.3821) = 18.8496 \pm 1.1463 = (17.7033, 19.9959).$$

The standard normal values at the specification limits are:

$$Z_1 = \frac{17.9 - 18.8496}{0.3821} = -2.485 \simeq -2.49$$

$$Z_2 = \frac{19.1 - 18.8496}{0.3821} = 0.655 \simeq 0.66.$$

The proportion below the LSL is 0.0064, while the proportion above the USL is 0.2546. This yields a total nonconforming proportion of 0.2610.

$$\hat{C}_p = \frac{19.1 - 17.9}{6(0.3821)} = 0.5234.$$

For a target volume of 18.5, we have:

$$\delta = \frac{18.8496 - 18.5}{0.3821} = 0.915.$$

$$\hat{C}_{pm} = \frac{0.5234}{\sqrt{1+(0.915)^2}} = 0.386.$$

To test $H_o : C_p \geq 0.6$ *vs.* $H_a : C_p < 0.6$, we have a one-sided test with $\alpha = 0.05$. A one-sided upper confidence limit is given by:

$$UCL = 0.386 \sqrt{\frac{\chi^2_{0.05,4}}{4}} = 0.386\sqrt{\frac{9.49}{4}} = 0.594.$$

Since the hypothesized value of $C_p = 0.6 > UCL = 0.594$, we reject H_o and conclude that the C_p index is less than 0.6.

9-37. From the tolerances on the length (ℓ) we have, $\mu_\ell = 4$ and $\sigma_\ell = 0.06/3 = 0.02$. From the tolerances on the width (w) we have $\mu_w = 5$ and $\sigma_w = 0.09/3 = 0.03$. The surface area of the solar cell is given by:

A = ℓw.

So, $\mu_A = 4(5) = 20$ cm^2.

Var (A) = $\sigma_A^2 = (\mu_w)^2 \, \sigma_\ell^2 + (\mu_\ell)^2 \, \sigma_w^2$

$= 5^2 (0.02)^2 + 4^2 (0.03)^2$

$= 0.0244.$

Hence, $\sigma_A = \sqrt{0.0244} = 0.1562.$

The natural tolerance limits on the surface area are:

$20 \pm 3(0.1562) = 20 \pm 0.4686 = (19.5314, 20.4686).$

The standard normal values at the specification limits are:

$$Z_1 = \frac{19.6 - 20}{0.1562} = -2.56$$

$$Z_2 = \frac{20.4 - 20}{0.1562} = 2.56.$$

The proportion below the LSL is 0.0052, which is the same as the proportion above the USL. So, the total proportion of nonconforming cells is 0.0104.

$$\hat{C}_p = \frac{20.4 - 19.6}{6(0.1562)} = 0.8536.$$

If the distribution of the surface area is not known, one approach could be to get some samples of solar cells and obtain empirical data on the area. Next, one could explore fitting of some known distributions (normal, exponential, gamma, Weibull) or to use some common transformations (power transformations of the Box-Cox type) to determine if the transformed data conforms to normality. Depending on the identified distribution, the process spread (assuming a 99.73% coverage, say) could be determined, which could then be used for calculating a capability index.

Alternatively, if no distribution is identifiable that best fits the data, a nonparametric approach may be taken based on the empirical observations. The 99.865[th] and 0.135[th] percentiles may be estimated from the empirical distribution, which could then be used to estimate the process spread. In this case, the nonparametric capability index, C_{pq}, could be used.

9-38. From the natural tolerances on weight (w), we have $\mu_w = 60$ and $\sigma_w = 5/3 = 1.667$. From the natural tolerances on height (h), we have $\mu_h = 1.7$ and $\sigma_h = 0.09/3 = 0.03$. The body mass index (BMI) represented by B, is given by:

$$B = \frac{w}{h^2}$$

$$\text{So, } \mu_B = \frac{60}{1.7^2} = 20.76.$$

$$\text{Var (B)} = \sigma_B^2 = \frac{(1.667)^2}{(1.7)^2} + \left(-\frac{2(60)}{1.7^3}\right)^2 (0.03)^2$$

$$= 1.4985.$$

$$\text{Hence, } \sigma_B = \sqrt{1.4985} = 1.2241.$$

The natural tolerance limits on BMI for this group of patients are:

$$20.76 \pm 3(1.2241) = 20.76 \pm 3.6723 = (17.0877, 24.4323).$$

The standard normal value at the obesity bound is:

$$Z = \frac{30 - 20.76}{1.2241} = 7.548.$$

Hence, the expected proportion of patients, from this group, above the bound of 30 is negligible (0.000).

An upper capability index is estimated as:

$$\hat{C}_p = \frac{30 - 20.76}{3(1.2241)} = 2.516.$$

A 95% lower confidence bound on C_p is:

$$\text{LCL} = 2.516 \sqrt{\frac{\chi^2_{0.95,19}}{19}} = 2.516 \sqrt{\frac{10.12}{19}} = 1.836.$$

9-39. We have $(1 - \alpha) = 0.99$ and $\gamma = 0.95$. The required sample size is:

$$n = 0.5 + \left(\frac{2 - 0.01}{0.01} \right) \frac{9.49}{4} = 472.63 \simeq 473.$$

Choose a sample of size 473 and rank the data values. The interval defined by the minimum and maximum values is the desired interval.

9-40. We have $(1 - \alpha) = 0.90$ and $\gamma = 0.95$. The sample size is given by:

$$n = \frac{\ell n(0.05)}{\ell n(0.90)} = 28.433 \simeq 29.$$

Choose a sample of size 29 and rank the data values. The range specifying the minimum value and above will define a one-sided lower nonparametric tolerance limit.

TABLE 10-8. Revised Probability Distribution Using Sample Information

Proportion nonconforming p	Prior Probability P(p)	P(X = 4 | p)	P(X = 4 and p)	Revised P(p)
0.05	0.20	0.1781	0.035620	0.5036
0.07	0.25	0.0887	0.022175	0.3135
0.09	0.35	0.0301	0.010535	0.1490
0.11	0.30	0.0080	0.002400	0.0339
			P(X = 4) = 0.07073	

For the third plan of n = 129, c = 4, for p = 0.01, np = 1.29, yielding P_a = 0.9893. So, of the three plans, the plan n = 129, c = 4 has the highest probability of acceptance for p = 0.01 and may be chosen.

10-26. It is given that P_a = 0.96 for p = 0.008. For c = 1, np = 0.312, which yields n = 0.312/0.008 = 39.

For c = 3, np = 1.267, yielding n = 1.267/0.008 = 158.37 ≈ 159.

For c = 4, np = 1.85, yielding n = 1.85/0.008 = 231.25 ≈ 232.

We desire P_a to be less than or equal to 0.04 when p = 0.05. Let us determine P_a for each of the three plans when p = 0.05.

For the plan c = 1, n = 39, for p = 0.05, np = 1.95, yielding P_a = 0.420. For the plan c = 3, n = 159, for p = 0.05, np = 7.95, yielding P_a = 0.0437.

For the plan c = 4, n = 232, for p = 0.05, np = 11.6, yielding P_a = 0.0122. So, of the three plans, the plan c = 4, n = 232 is preferable and meets the criterion.

10-27. We have α = 0.05, p_1 = AQL = 0.009, β = 0.10, p_2 = LQL = 0.065. We get p_2/p_1 = 0.065/0.009 = 7.22, yielding candidates c = 1 or c = 2. For satisfying the consumer's stipulation, the plans are:

c = 1, np_2 = 3.890, yielding n = 3.890/0.065 = 59.85 ≈ 60.

c = 2, np_2 = 5.322, yielding n = 5.322/0.065 = 81.88 ≈ 82.

Let us find how close these plans come to satisfying the producer's stipulation. For the plan c = 1, n = 60, np_1 = 0.355, yielding p_1 = 0.355/60 = 0.00592. For the plan c = 2, n = 82, np_1 = 0.818, yielding p_1 = 0.818/82 = 0.00997. The plan c = 2, n = 82 comes closest to meeting the producer's stipulation, since p_1 = 0.00997 is closer to the desired value of 0.009. This plan will also be slightly more stringent than the requirements.

10-28. It is given that $\alpha = 0.05$, $p_1 = AQL = 0.013$, $\beta = 0.10$, $p_2 = LQL = 0.071$. We have $p_2/p_1 = 0.071/0.013 = 5.46$, yielding candidates $c = 2$ or $c = 3$. For satisfying the producer's stipulation, the plans are:

$c = 2$, $np_1 = 0.818$, yielding $n = 0.818/0.013 = 62.92 \simeq 63$.

$c = 3$, $np_1 = 1.366$, yielding $n = 1.366/0.013 = 105.08 \simeq 106$.

Let us find how close these plans come to satisfying the consumer's stipulation. For the plan $c = 2$, $n = 63$, $np_2 = 5.322$, yielding $p_2 = 5.322/63 = 0.0845$. For the plan $c = 3$, $n = 106$, $np_2 = 6.681$, yielding $p_2 = 6.681/106 = 0.0630$. The plan $c = 3$, $n = 106$ comes closest to meeting the consumer's stipulation, since $p_2 = 0.0630$ is closer to the desired value of 0.071, and is also slightly more stringent than the requirements.

10-29. It is given that $\alpha = 0.05$, $p_1 = AQL = 0.02$, $\beta = 0.10$, $p_2 = LQL = 0.07$. We get $p_2/p_1 = 0.07/0.02 = 3.5$, yielding candidates $c = 5$ or $c = 6$. For satisfying the producer's stipulation exactly, the plans are:

$c = 5$, $np_1 = 2.613$, yielding $n = 2.613/0.02 = 130.65 \simeq 131$.

$c = 6$, $np_1 = 3.286$, yielding $n = 3.286/0.02 = 164.3 \simeq 165$.

For satisfying the consumer's stipulation exactly, the plans are:

$c = 5$, $np_2 = 9.274$, yielding $n = 9.274/0.07 = 132.48 \simeq 133$.

$c = 6$, $np_2 = 10.532$, yielding $n = 10.532/0.07 = 150.46 \simeq 151$.

Thus, the plan with the largest sample size is $c = 6$, $n = 165$. The plan with the smallest sample size is $c = 5$, $n = 131$.

10-30. $P_{a1} = P(x_1 \leq 1 \mid n_1p = 2) = 0.406$.

$P_{a2} = P(x_1 = 2 \mid n_1p = 2)\, P(x_2 \leq 3 \mid n_2p = 4.4) + P(x_1 = 3 \mid n_1p = 2)\, P(x_2 \leq 2 \mid n_2p = 4.4)$

$= (0.677 - 0.406)(0.3602) + (0.857 - 0.677)(0.1868)$

$= 0.1312$.

So, probability of lot acceptance $= 0.406 + 0.1312 = 0.5372$.

P(rejecting lot on first sample) $= P(x_1 \geq 4 \mid n_1p = 2) = 1 - 0.857 = 0.143$.

10-31. $P_{a1} = P(x_1 = 0 \mid n_1p = 1.8) = 0.165$.

$P_{a2} = P(x_1 = 1 \mid n_1p = 1.8)\, P(x_2 \leq 5 \mid n_2p = 3) + P(x_1 = 2 \mid n_1p = 1.8)\, P(x_2 \leq 4 \mid n_2p = 3)$

$+ P(x_1 = 3 \mid n_1p = 1.8) P(x_2 \leq 3 \mid n_2p = 3) + P(x_1 = 4 \mid n_1p = 1.8) P(x_2 \leq 2 \mid n_2p = 3)$

$= (0.298)(0.916) + (0.268)(0.815) + (0.160)(0.647) + (0.073)(0.423) = 0.6258.$

So, probability of lot acceptance $= 0.165 + 0.6258 = 0.7908$. Probability of accepting a lot on the first sample $= 0.165$. P(rejecting a lot on first sample) $= P(x_1 \geq 5 \mid n_1p = 1.8) = 1 - 0.964 = 0.036$. Hence, probability of making a decision on the first sample $= 0.165 + 0.036 = 0.201$.

10-32. $P_{a1} = P(x_1 \leq 1 \mid n_1p = 1.0) = 0.736.$
$P_{r1} = P(x_1 \geq 4 \mid n_1p = 1.0) = 1 - 0.981 = 0.019$, so $P_1 = 0.736 + 0.019 = 0.755.$

$ASN = 50 + 110(1 - 0.755) = 76.95.$

$P_{a2} = P(x_1 = 2 \mid n_1p = 1.0) P(x_2 \leq 3 \mid n_2p = 2.2) + P(x_1 = 3 \mid n_1p = 1.0) P(x_2 \leq 2 \mid n_2p = 2.2)$

$= (0.184)(0.819) + (0.061)(0.623) = 0.1887.$

$ATI = 50(0.736) + (50 + 110)(0.1887) + 1200 (1 - 0.736 - 0.1887) = 157.352.$

10-33. It is given that $\alpha = 0.05$, $p_1 = AQL = 0.018$, $\beta = 0.10$, $p_2 = LQL = 0.085$, $N = 1500$, and $n_1 = n_2$. $R = p_2/p_1 = 0.085/0.018 = 4.72$. So, $c_1 = 2$ and $c_2 = 4$. Now $n_1 = 1.16/0.018 = 64.4 \simeq 65$. Hence, the double sampling plan is: $n_1 = 65$, $c_1 = 2$, $r_1 = 5$, $n_2 = 65$, $c_2 = 4$, $r_2 = 5$.

For $p_1 = 0.018$ for which $P_a = 0.95$, $ASN/n_1 = 1.105$, or $ASN = 1.105(65) = 71.825$.

10-34. It is given that $\alpha = 0.05$, $p_1 = 0.023$, $\beta = 0.10$, $p_2 = 0.095$, $N = 2000$, and $n_2 = 2n_1$. $R = p_2/p_1 = 0.095/0.023 = 4.13$. So, $c_1 = 1$ and $c_2 = 5$. Now $n_1 = 4.01/0.095 = 42.3 \simeq 43$. Hence, the double sampling plan is: $n_1 = 43$, $c_1 = 1$, $r_1 = 6$, $n_2 = 86$, $c_2 = 5$, $r_2 = 6$.

For $p_1 = 0.023$ for which $P_a = 0.95$, $ASN/n_1 = 1.498$, or $ASN = 1.498(43) = 64.414$.

10-35. It is given that $\alpha = 0.05$, $p_1 = 0.018$, $\beta = .10$, $p_2 = 0.085$, $N = 1500$, and $n_2 = 2n_1$. $R = p_2/p_1 = 0.085/0.018 = 4.72$. So, $c_1 = 1$ and $c_2 = 4$. Now $n_1 = 3.92/0.085 = 46.1 \simeq 47$. Hence, the double sampling plan is: $n_1 = 47$, $c_1 = 1$, $r_1 = 5$, $n_2 = 94$, $c_2 = 4$, $r_2 = 5$.

10-36. It is desired that $P_a = 0.5$ for $p = 0.05$. $R = p_2/p_1 = 0.085/0.018 = 4.72$. So, $c_1 = 2$ and $c_2 = 4$. Now $n_1 = 2.90/0.05 = 58$. Hence, the double sampling plan is: $n_1 = 58$, $c_1 = 2$, $r_1 = 5$, $n_2 = 58$, $c_2 = 4$, $r_2 = 5$.

10-37. The sampling plan is n = 105, c = 2. The AOQL for the plan is 1.2%, meaning that the worst quality, on average, leaving the inspection station will not exceed 1.2% nonconforming.

10-38. Since the process average is not specified, in being conservative and using the last column of the process average values, the sampling plan is n = 370, c = 13. The AOQL for the plan is 2.1%, meaning that the worst quality, on average, leaving the inspection station will not exceed 2.1% nonconforming.

10-39. The sampling plan is n = 42, c = 2. The LQL is 12.4%, meaning that lots with a nonconformance rate of 12.4% have a 10% chance of being accepted by the selected plan.

10-40. It is given that n = 5, i = 3, p = 0.06.

$$P(0,5) = \frac{5!}{0!5!} (0.06)^0 (0.94)^5 = 0.7339$$

$$P(1,5) = \frac{5!}{1!4!} (0.06)^1 (0.94)^4 = 0.2342.$$

Hence, $P_a = 0.7339 + (0.2342)(0.7339)^3 = 0.8265$.

10-41. The first opportunity to reject is on the second item inspected. The first opportunity to accept is on the 16th item inspected.

10-42. It is given that $\alpha = 0.05$, $p_1 = 0.008$, $\beta = 0.07$, and $p_2 = 0.082$. The parameters of the acceptance and rejection lines are calculated as follows:

$$k = \ell n \left(\frac{0.0821(1 - 0.008)}{0.008(1 - 0.082)} \right) = 2.4048$$

$$h_a = \ell n \left(\frac{1 - 0.05}{0.07} \right) / 2.4048 = 1.084$$

$$h_r = \ell n \left(\frac{1 - 0.07}{0.05} \right) / 2.4048 = 1.215$$

$$s = \ell n \left(\frac{1 - 0.008}{1 - 0.082} \right) / 2.4048 = 0.032.$$

The acceptance line is: $d_a = -1.084 + (0.032)n$.

The rejection line is: $d_r = 1.215 + (0.032)n$.

The first opportunity to reject is on the 2nd item inspected. The first opportunity to accept is on the 34th item inspected.

10-43. It is given that $k_1 = 0.50$, $k_2 = 225$, and $p = 0.003$. The ratio $k_1/k_2 = 0.50/225 = 0.0022 < p = 0.003$. So, the policy calls for 100% inspection.

10-44. It is given that $k_1 = 1.00$, $k_2 = 225$, and $p = 0.003$. The ratio $k_1/k_2 = 1.00/225 = 0.0044 > p = 0.003$. So, the policy calls for no inspection.

10-45. Since the kp rule calls for 100% inspection, the monthly cost would be $0.50 x 3000. With no inspection, the replacement cost would be $225 for a proportion of 0.003 of the product items. This replacement cost would be $225 x 0.003 x 3000 = $0.675 x 3000. Therefore, on average, the savings in total inspection costs is $0.175 per item, for a monthly savings of $525.

10-46. Since the kp rule calls for no inspection, the monthly cost, on average, would be for replacement of nonconforming units, which is $225 x 0.003 x 2000 = $0.675 x 2000. The cost of 100% inspection on a monthly basis is $1.00 x 2000. Therefore, on average, the monthly savings is $0.325 x 2000 = $650.

10-47. It is given that $k_1 = 0.20$, $k_2 = 50$, and $p = 0.005$. The ratio $k_1/k_2 = 0.20/50 = 0.004 < p = 0.005$. So, the policy calls for 100% inspection. For 100% inspection, the total inspection cost per unit is $0.20. With no inspection, the total inspection cost will be associated with repair and replacement of nonconforming units. On average, this cost per unit will be $50 x 0.005 = $0.25. Therefore, by using 100% inspection, the average savings in inspection costs per unit is $0.05.

10-48. It is given that USL = 30, $\sigma = 5$, $\alpha = 0.05$, $\beta = 0.08$, $\overline{X}_1 = 30 - (2.3)5 = 18.5$, $\overline{X}_2 = 30 - 5 = 25$. We have: $Z_\alpha = Z_{.05} = 1.645$, $Z_\beta = Z_{.08} = -1.405$. The sample size is:

$$n = \left(\frac{(-1.405 - 1.645)5}{18.5 - 25} \right)^2 = 5.50 \approx = 6.$$

The acceptance limit is:

$$\overline{X}_a = \frac{(-1.405)(18.5) - (1.645)(25)}{-1.405 - 1.645} = 22.006.$$

The plan works as follows. A random sample of size 6 is chosen and the sample average resistance is found. If the sample average is greater than 22.006, the lot is rejected; otherwise, it is accepted.

10-49. It is given that LSL = 25, $\alpha = 0.06$, $\beta = 0.07$, $Z_\alpha = -1.555$, $Z_\beta = 1.476$, and $\sigma = \sqrt{6} = 2.449$. We have:

$$\overline{X}_1 = 25 + 1.88(2.449) = 29.604.$$

$$\overline{X}_2 = 25 + 1.41(2.449) = 28.453.$$

The sample size is:

$$n = \left(\frac{(1.476 + 1.555)2.449}{29.604 - 28.453} \right)^2 = 41.59 \simeq 42.$$

The acceptance limit is:

$$\overline{X}_a = \frac{1.476(29.604) - (-1.555)28.453}{1.476 - (-1.555)} = 29.0135.$$

The plan works as follows. A random sample of size 42 is chosen and the sample average breaking strength is found. If the sample average is less than 29.0135, the lot is rejected; otherwise, it is accepted.

10-50. We have $\overline{X}_{2L} = 800$, $\overline{X}_{2U} = 1200$, $\overline{X}_1 = 1000$, $\beta = 0.08$, $\alpha = 0.04$, $\sigma = 80$, $Z_{\alpha/2} = 2.054$, $Z_\beta = 1.405$. So,

$$2.054 = (\overline{X}_{Ua} - 1000)/(80/\sqrt{n})$$

$$-2.054 = (\overline{X}_{La} - 1000)/(80/\sqrt{n})$$

$$1.405 = (\overline{X}_{La} - 800)/(80/\sqrt{n})$$

$$-1.405 = (\overline{X}_{Ua} - 1200)/(80/\sqrt{n}).$$

From the above equations, $\overline{X}_{La} + \overline{X}_{Ua} = 2000$. Also,

$$\sqrt{n} = \frac{(2.054 + 1.405)80}{(1000 - 800)} = 1.3836, \text{ yielding } n = 2.$$

So, $\overline{X}_{La} = 1000 - (2.054)(80/\sqrt{2}) = 883.808$, and $\overline{X}_{Ua} = 1116.192$. The plan operates as follows. A random sample of size 2 is chosen from the lot and the average tensile strength is computed. If the sample average is less than 883.808 or greater than 1116.192, the lot is rejected. Otherwise, the lot is accepted.

10-51. We have $U = 45$, $L = 40$, $\beta = 0.06$, $\alpha = 0.06$, $\sigma = 0.8$, $Z_{\alpha/2} = 1.88$, $Z_\beta = 1.555$. From the information given:

$$\overline{X}_{2L} = 40 + 1.405(0.8) = 41.124$$

$$\overline{X}_{2U} = 45 - 1.405(0.8) = 43.876, \text{ and } \overline{X}_1 = 42.5.$$

So, $1.88 = (\overline{X}_{Ua} - 42.5)/(0.8/\sqrt{n})$

$-1.88 = (\overline{X}_{La} - 42.5)/(0.8/\sqrt{n})$

$1.555 = (\overline{X}_{La} - 41.124)/(0.8/\sqrt{n})$

$-1.555 = (\overline{X}_{Ua} - 43.876)/(0.8/\sqrt{n}).$

From the above equations, $\overline{X}_{La} + \overline{X}_{Ua} = 85$. Also,

$$\sqrt{n} = \frac{(1.88 + 1.555)(0.8)}{42.5 - 41.124} = 1.997, \text{ yielding } n = 4.$$

So, $\overline{X}_{La} = 42.5 - (1.88)(0.8/\sqrt{4}) = 41.748$ and $\overline{X}_{Ua} = 43.252$. The plan operates as follows. A random sample of size 4 is chosen from the lot and the average length is computed. If the sample average is less than 41.748 or greater than 43.252, the lot is rejected. Otherwise, the lot is accepted.

10-52. It is given that $\overline{X}_1 = 0.15$, $\alpha = 0.05$, $\overline{X}_2 = 0.34$, $\beta = 0.20$, and $\hat{\sigma} = 0.25$. The parameter $\lambda = |0.15 - 0.34|/0.25 = 0.76$. The sample size is approximately 15. The t-value corresponding to an upper tail area of 0.05 with 14 degrees of freedom is 1.761. The plan works as follows. A random sample of size 15 is chosen and the sample mean and standard deviation are computed. The following statistic is computed: $t = (\overline{X} - 0.15)/(s/\sqrt{15})$. If $t > 1.761$, the lot is rejected. Otherwise, the lot is accepted.

10-53. It is given that $\alpha = 0.05$, $\beta = 0.20$, $\hat{\sigma} = 0.25$:

$$\overline{X}_1 = 0.30 - 0.25 = 0.05$$

$$\overline{X}_2 = 0.30 + 0.25(0.8) = 0.50.$$

The parameter $\lambda = |\,0.05 - 0.50\,|/0.25 = 1.8$. The sample size is approximately 4. The t-value corresponding to an upper tail area of 0.05 with 3 degrees of freedom is 2.353. The plan works as follows. A random sample of size 4 is chosen and the sample mean and standard deviation are computed. The following statistic is computed: $t = (\bar{X} - 0.05)/(s/\sqrt{4})$. If $t > 2.353$, lot is rejected. Otherwise, the lot is accepted.

10-54. It is given that L = 89, $\hat{\sigma} = 4$, $\bar{X}_1 = 94$, $\alpha = 0.05$, $\bar{X}_2 = 86$, $\beta = 0.15$. The parameter $\lambda = |\,94 - 86\,|/4 = 2$. The sample size is approximately 4. The t-value corresponding to a lower tail area of 0.05 with 3 degrees of freedom is –2.353. The plan works as follows. A random sample of size 4 is chosen and the sample mean and standard deviation are computed. The following statistic is computed: $t = (\bar{X} - 94)/(s/\sqrt{4})$. If $t < -2.353$, the lot is rejected. Otherwise, the lot is accepted.

10-55. It is given that U = 4, $\hat{\sigma} = 0.05$, $\alpha = 0.05$, $p_1 = 0.01$, $\beta = 0.08$, $p_2 = 0.07$, $Z_\alpha = 1.645$, $Z_1 = 2.33$, $Z_\beta = 1.405$, $Z_2 = 1.476$. The sample size is:

$$n = \left(\frac{1.645 + 1.405}{2.33 - 1.476}\right)^2 = 12.755 \simeq 13.$$

In order to satisfy the value of $\alpha = 0.05$, we have:

$$k = 2.33 - 1.645/\sqrt{13} = 1.8737.$$

We first demonstrate using Form 1: $Z_U = (4 - 3.05)/0.5 = 1.9$. Since $Z_U = 1.9 > k = 1.8737$, the lot is accepted. Using Form 2: $Q_U = 1.9\sqrt{13/12} = 1.977 \simeq 1.98$. The estimated proportion of nonconforming items is 0.0239. Now, $k\sqrt{n/(n-1)} = 1.8737\sqrt{13/12} = 1.95$. The area above this standard normal value is 0.0256 which is M, the maximum allowable proportion nonconforming. Since $0.0239 < M = 0.0256$, the lot is accepted.

10-56. It is given that U = 0.015, $\hat{\sigma} = 0.0014$, $\beta = 0.05$, $p_2 = 0.11$, $\alpha = 0.04$, $p_1 = 0.02$, $Z_\alpha = 1.75$, $Z_\beta = 1.645$, $Z_1 = 2.055$, $Z_2 = 1.226$. The sample size is:

$$n = \left(\frac{1.75 + 1.645}{2.055 - 1.226}\right)^2 = 16.77 \simeq 17.$$

In order to satisfy the value of $\beta = 0.05$, we have:

$$k = 1.226 + 1.645/\sqrt{17} = 1.625.$$

We first demonstrate using Form 1: $Z_U = (0.015 - 0.013)/0.0014 = 1.429$. Since $Z = 1.429 < k = 1.625$, the lot is rejected. Using Form 2: $Q_U = 1.429\sqrt{17/16} = 1.473$, the estimated proportion nonconforming is 0.07038. Also, $k\sqrt{17/16} = 1.625\sqrt{17/16} = 1.675$, yielding $M = 0.0470$. Since $0.07038 > M = 0.0470$, the lot is rejected.

10-57. It is given that $L = 25$, $\hat{\sigma} = 0.03$, $p_1 = 0.015$, $\alpha = 0.08$, $p_2 = 0.08$, $\beta = 0.12$, $Z_\alpha = 1.405$, $Z_1 = 2.17$, $Z_\beta = 1.175$, $Z_2 = 1.405$. The sample size is:

$$n = \left(\frac{1.405 + 1.175}{2.17 - 1.405}\right)^2 = 11.37 \approx 12.$$

In order to satisfy the value of $\alpha = 0.08$, we have:

$$k = 2.17 - 1.405/\sqrt{12} = 1.7644.$$

In order to satisfy the value of $\beta = 0.12$, we have:

$$k = 1.405 + 1.175/\sqrt{12} = 1.7442.$$

CHAPTER 11

RELIABILITY

11-1. Reliability is the probability of a product performing its intended function for a stated period of time under certain specified conditions. Since customer satisfaction plays a fundamental role in quality control and improvement, such is influenced by the functional performance of the product over time. The expected operational time is an extended period, usually much beyond the warranty time. Ensuring high product reliability will therefore ensure greater satisfaction with the product.

11-2. The typical life cycle of a product is represented by the bathtub curve, which consists of a debugging phase, a chance-failure phase, and a wear-out phase. In the debugging phase, a drop in the failure rate is observed with time as initial problems are addressed during prototype testing. In the chance-failure phase, failures occur randomly and independently where the failure rate remains constant. This phase typically represents the useful life of the product. In the wear-out phase, an increase in the failure rate is observed with time as parts age and wear out. For the debugging phase and the wear-out phase, a Weibull or gamma distribution may be used. Through adequate selection of the parameters of such a distribution, both increases and decreases in failure rate over time can be modeled. For the chance-failure phase, an exponential distribution, that has a constant failure rate, can be used.

11-3. Reliability of a system could be improved through an increase in the reliability of components that make up the system through adequate consideration in the design stage. However, this approach may not be very effective for systems with a large number of components in series. In such cases, system reliability can be improved by placing components in parallel or by having standby components in parallel. For components in parallel, even though they are redundant, all components are assumed to operate simultaneously. On the other hand, for standby components in parallel, only one component in the parallel configuration is operating at any given time. Thus, standby components wait to take over operation upon failure of the currently operating component.

The availability of a system, at a specified time t, is the probability that the system will be operating at time t. The availability index is defined as the ratio of the operating time to the total time (which consists of the operating time plus the downtime). Availability may be increased by reducing the downtime through preventive or condition-based maintenance. Through an adequate design, the operational time could be improved, which will increase availability. Reducing the time to repair a system, through improved methods, will also increase availability.

11-4. In a failure-terminated testing plan, the tests are terminated when a preassigned number of failures occurs in the chosen sample. Lot acceptance is based on the accumulated test times of the items. In a time-terminated testing plan, the test is terminated when a preassigned time is reached. Lot acceptance is based on the observed number of failures during the test time. In a sequential test for reliability, no prior decision is made on the number of failures or the time to conduct the test. The accumulated results of the test are used to determine whether to accept the lot, reject the lot, or continue testing. The cumulative number of failures is plotted versus the accumulated test time of items. The acceptance and rejection lines are found based on a chosen level of producer's risk and acceptable mean life, and a chosen level of consumer's risk and a minimum mean life.

11-5.   Reliability = exp (-0.00006)(4000) = 0.7866. MTTF = $1/\lambda$ = 1/0.00006 = 16666.667 hrs.

Availability = $\dfrac{\mu}{\lambda + \mu}$

$$= \dfrac{0.004}{0.00406} = 0.985.$$

11-6.   R(t) = exp (- λ (6000)) = 0.92. This yields λ = 0.13896 x 10^{-4}/hour, or MTTF = $1/\lambda$ = 71958.314.

11-7.   It is given that α = 300 and β = 0.5. R(t) = exp [-(500/300)$^{0.5}$] = 0.275. MTTF = 300 x Γ (1/0.5 + 1) = 600.

11-8.   Reliability of the remote control unit = $(0.9994)^{40}$ = 0.9763. Reliability of redesigned unit = $(0.9994)^{25}$ = 0.9851.

11-9.   We have, 0.996 = exp [-λ_s (3000)], which yields λ_s = 0.1336 x 10^{-5}/hour. The failure rate for each component is λ = λ_s /25 = 0.1336 x 10^{-5}/25 = 0.53 x 10^{-7}/hour or 5.3/10^8 hours. MTTF for each component is $1/\lambda$ = 18.7125 x 10^6 hours.

11-10. System reliability is R_s = 1 – (1 – 0.93)(1 – 0.88)(1 – 0.95)(1 – 0.92) = 0.9999664.

11-11. For each component, MTTF = 3000 hours, yielding an individual failure rate λ = 1/3000 = 3.333 x 10^{-4}/hour. Reliability of the subassembly is R_s = 1 – [1 – exp (-(1/3000)2500)]^4 = 0.8978. For the subassembly, MTTF = 3000 [1 + 1/2 + 1/3 + 1/4] = 6250 hours. If the MTTF of the subassembly were to be 6600, we have 6600 = $(1/\lambda)$[1 + 1/2 + 1/3 + 1/4], yielding λ = 3.156 x 10^{-4}/hour. So, for each component, MTTF = 3168 hours.

11-12. The reliability of the subsystem with components A and B is R_1 = 1 – (1 – 0.96)(1 – 0.92) = 0.9968. The reliability of the subsystem with components E, F, and G is R_2 = 1 – (1 – (0.95)(0.88))(1 – 0.90) = 0.9836. So, the system reliability is R_s = (0.9968)(0.94)(0.89) (0.9836) = 0.8202. To improve system reliability, try to improve the reliabilities of C and D.

11-13. For the subsystem with components A and B, MTTF = (1/0.0005)(1 + 1/2) = 3000. For the subsystem with components E, F, and G, MTTF = (1/0.0064)(1 + 1/2) = 234.375. Note that the failure rate of the subsystem with components E and F in series is 0.0004 + 0.006 = 0.0064. Now, the system failure rate is:

λ_s = 1/3000 + 0.0003 + 0.0008 + 1/234.375 = 0.0057/hour, yielding a mean time to failure for the system $= 1/0.0057 = 175.4386$ hours. The reliability of the system after 1000 hours is $R_s = \exp(-0.0057(1000)) = 0.003346$.

11-14. Reliability of system is:

$$R_s = \exp[-0.008(400)][1 + 0.008(400) + (0.008(400))^2/2$$

$$+ (0.008(400))^3/6 + (0.008(400))^4/24]$$

$$= 0.7806.$$

$\text{MTTF} = (n + 1)/\lambda = 5/0.008 = 625$ hours.

If all five units were operating in parallel, system reliability would be:

$$R_s = 1 - [1 - \exp(-0.008(400))]^5$$

$$= 0.18786.$$

In that case, $\text{MTTF} = (1/0.008)(1 + 1/2 + 1/3 + 1/4 + 1/5)$

$$= 285.4167 \text{ hours.}$$

11-15. For the subsystem consisting of components A and B, $\text{MTTF} = (1/0.0005)(2) = 4000$. The reliability of this subsystem is:

$$R_1 = \exp(-0.0005(1000))[1 + 0.0005(1000)]$$

$$= 0.9098.$$

The reliability of the subsystem consisting of components C and D is:

$$R_2 = \exp(-0.0011(1000)) = 0.33287.$$

The reliability of the subsystem consisting of components E, F, and G is:

$$R_3 = \exp(-(1/234.375)(1000)) = 0.0140.$$

So, the system reliability is $R_s = (0.9098)(0.33287)(0.0140) = 0.00424$. The system failure rate is:

$\lambda_s = 1/4000 + 0.0011 + 1/234.375 = 0.005617/\text{hour}$, yielding an $\text{MTTF} = 1/0.005617 = 178.041$ hours.

11-16. The computations for the OC curve are shown in Table 11-1.

TABLE 11-1. Computations for OC Curve

Mean Life θ	Failure rate $\lambda = 1/\theta$	nT λ	P_a
500	0.002	10.8	0.0068
800	0.00125	6.75	0.097
900	0.00111	6.0	0.151
1000	0.001	5.4	0.2146
2000	0.0005	2.7	0.174
3000	0.00033	1.8	0.891
4000	0.00025	1.35	0.9515
5000	0.0002	1.08	0.9754
6000	0.000167	0.9	0.987
8000	0.000125	0.675	0.9947

For a producer's risk of 0.05, the associated quality level of batches as indicated by their mean life is about 4000 hours. For a consumer's risk is 0.10, the associated quality level of batches has a mean life of about 800 hours.

11-17. The accumulated life for the test units is $T_6 = (530 + 590 + 670 + 700 + 720 + 780) + (20 - 6)\ 780 = 14910$ hours. Mean time to failure is estimated as 14910/6 = 2485 hours. The estimated failure rate is 1/2485 = 0.000402/hour. A 95% confidence interval for the mean life is:

$$\frac{2(14910)}{26.12} < \theta < \frac{2(14910)}{5.63}$$

or $1141.654 < \theta < 5296.625$.

11-18. The accumulated life on the test units is $T_6 = 20(780) = 15600$ hours. Mean time to failure is estimated as 15600/6 = 2600 hours. The estimated failure rate is 1/2600 = 0.000385/hour. A 90% confidence interval for the mean life is:

$$\frac{2(15600)}{23.68} < \theta < \frac{2(15600)}{6.57}$$

or $1317.568 < \theta < 4748.858$.

11-19. The accumulated life on the test units is $T_5 = (610 + 630 + 680 + 700 + 720) + (25 - 5)\ 800 = 19340$ hours. Mean life is estimated as 19340/5 = 3868 hours. The estimated failure rate is 1/3868 = 0.0002585/hour. A 95% confidence interval for the mean life is:

$$\frac{2(19340)}{23.34} < \theta < \frac{2(19340)}{4.40}$$

or $1657.241 < \theta < 8790.909$.

11-20. It is given that r = 8, n = 15, θ_0 = 900, α = 0.10. An estimate of the mean life is $\hat{\theta}$ = [400 + 430 + 435 + 460 + 460 + 490 + 520 + 530 + (15 − 8)530]/8 = 7435/8 = 929.375 hours. From the standard, the code is C-8, and C/θ_0 = 0.582. The acceptability criterion C is 900(0.582) = 523.8. Since the estimated mean life exceeds 523.8, the lot should be accepted.

11-21. Using the information from Problem 11-20, an estimate of the mean life is $\hat{\theta}$ = 15(530)/8 = 993.75 hours. From the standard, the code is C-8, and C/θ_0 = 0.582. The acceptability criterion C is 900(0.582) = 523.8. Since the estimated mean life exceeds 523.8, the lot should be accepted.

11-22. It is given that r = 3, n = 8, θ_0 = 600, and α = 0.01. An estimate of the mean life is $\hat{\theta}$ = [200 + 240 + 250 + (8 - 3)250]/3 = 1940/3 = 646.667 hours. From the standard, the code is A-3, and C/θ_0 = 0.145. The acceptability criterion C is 600(0.145) = 87. Since the estimated mean life exceeds 87, the lot should be accepted.

11-23. It is given that θ_0 = 1500, α = 0.05, θ_1 = 600, and β = 0.10. We have θ_1/θ_0 = 600/1500 = 0.4, for which the code is B-11. This gives a rejection number of 15, a sample size of 75, and T/θ_0 = 0.136. So, the test time is T = 1500(0.136) = 204 hours.

11-24. Using the information from Problem 11-23, the code is B-11. This gives a rejection number of 15, a sample size of 75, and T/θ_0 = 0.123. So, the test time is T = 1500(0.123) = 184.5 hours.

11-25. It is given that θ_0 = 1400, α = 0.05, r = 7, and n = 35. For r = 7 and n = 5r, T/θ_0 = 0.103. The test termination time is T = 1400(0.103) = 144.2 hours.

11-26. Using the information from Problem 11-25, for r = 7 and n = 5r, T/θ_0 = 0.094. The test termination time is T = 1400(0.094) = 131.6 hours.

11-27. It is given that θ_0 = 6000, α = 0.01, θ_1 = 2000, β = 0.10, and T = 1200. We have: θ_1/θ_0 = 2000/6000 = 1/3, and T/θ_0 = 1200/6000 = 1/5. From the table, r = 13 and n = 30.

11-28. Using the information from Problem 11-27, for θ_1/θ_0 = 1/3 and T/θ_0 = 1/5, we get r = 13 and n = 37.

11-29. It is given that θ_0 = 7500, α = 0.10, θ_1 = 1500, β = 0.05, and T = 2500. We have: θ_1/θ_0 = 1500/7500 = 1/5, and T/θ_0 = 2500/7500 = 1/3. From the table, r = 4 and n = 5.

11-30. Using the information from Problem 11-29, for θ_1/θ_0 = 1/5 and T/θ_0 = 1/3, we get r = 4 and n = 7.

CHAPTER 12

EXPERIMENTAL DESIGN AND THE TAGUCHI METHOD

12-1. Suppose that patients are to be prescribed medication based on the criticality of their condition. Criticality could be defined on the basis of cholesterol and systolic blood pressure levels. Three levels of cholesterol (high, medium, and low) are defined based on chosen benchmarks. Similarly, two levels of systolic blood pressure (high and low) are defined based on chosen benchmarks. It is desired to study the impact of cholesterol and systolic blood pressure on recovery time (response variable). So, here there are two factors: cholesterol level and systolic blood pressure. The factor, cholesterol, has 3 levels, while the factor systolic blood pressure has 2 levels. A treatment is a combination of factor levels. Here, there are six treatments, corresponding to the 6 possible combinations of cholesterol level and systolic blood pressure level. The number of treatments is 6.

12-2. In a financial institution, it may be of interest to study the defaulting of loans associated with home mortgages. Thus, one wishes to identify key characteristics of borrowers, who will likely not default on their mortgages. Borrowers may be classified by income category (< $50K, $50 – $100K, $100 – $200K, > $200K), type of business (education and non-profit organizations, for-profit organizations), number of years of employment (0–5 years, 5 – 10 years, > 10 years), and credit rating (low, medium, high). Analysis through experimental design could indicate critical characteristics that will predict selection of applicants who will not default on their loans.

12-3. Replication involves the ability to obtain several observations under similar process conditions. This enables us to obtain an estimate of the experimental error (inherent variation in the process). In a semiconductor manufacturing company this could represent results on proportion of nonconforming wafers when process parameters (pressure, temperature, etch time) are kept constant.

Randomization of assignment of treatments to experimental units is necessary to eliminate bias. This means that through randomization, the impact of all other factors not being considered in the experiment, is averaged out. In a semiconductor manufacturing company, we could select the treatment combinations (pressure x temperature x etch time) and assign to run them in a random order, rather than in a systematic fashion (say low pressure to medium pressure to high pressure).

Blocking is the ability to control the impact of nuisance parameters on the response variable. While our objective is necessarily not focused on these parameters (we are interested in the design parameters), we would like to partition out their impact so that we may have a clear focus on the impact of the design parameters. In the situation considered here, blocking factors could be the vendor that provides components (say 3 vendors) or the manufacturing line (there could be 2 lines that make similar wafers). The variation due to the blocks is partitioned out from the error sum of the squares. The treatment sum of squares represents the variation, say, due to changes in the treatments: pressure x temperature x etch time. Through a reduction in the error sum of squares, by blocking, it may be possible to improve the sensitivity of the experiment to determine the significance of the treatments.

12-4. Interaction exists between two factors, when the nature of the relationship between one factor and the response variable is influenced by the level of the second factor. Suppose that the response variable represents satisfaction derived from movies, expressed on a 100-point scale. One factor could be the degree of action-theme (could have three levels, low, medium, and high) and a second factor could be age category (<35, ≥ 35). For the younger age category, the degree of satisfaction could increase with the degree of action-theme. On the other hand, for the older age category, the degree of satisfaction could initially increase from a low to medium degree of action-theme but could decrease thereafter as the degree of action-theme increases further. Hence, in this case we would say that there is a possible interaction between degree of action-theme and age.

12-5. A quantitative variable is expressed on a measurable scale. For example, number of tons of shipped goods or unloading time (in days) of a tanker. For quantitative variables, interpolation of the response variable is feasible. A qualitative variable is indicated categorically, for example the company that obtains a contract to ship goods, from 3 possible bidders, or the method of transport selected (rail, truck, or ship). Here interpolation of the response variable is not feasible.

12-6. In a fixed effects model, inferences on the means pertain only to the treatments represented in the experiment. In a random effects model, the treatments are a random selection from many that are possible, not all of which can possibly be represented in the experiment. In such a model, even though the inferences pertain to all of the possible treatments, they are not on the treatment means but rather on the treatment variance. In the logistics area, suppose the possible number of routings from a source to a destination (through transshipment at intermediate stations) is extremely large, such that it may not be feasible to study each routing. In a fixed effects model, we may select a limited number of routings (say, 10) and determine if there are significant differences in the mean transportation time for the 10 routings. In a random effects model, we may randomly select some limited number of routings (from the large number of possible routings) and determine if the variability of the transportation time approaches zero.

12-7. In a completely randomized design, the treatments are assigned to the experimental units in a random manner. In a gasoline refining process, assume that the treatments are combinations of pressure x temperature x amount of catalyst, with each factor having a certain number of levels. If the treatments are assigned randomly to determine impact on, say, octane rating, with no pre-defined systematic variation in any of the factors, a completely randomized design exists.

However, there could be some other factors, not considered in the experiment, that could influence octane rating. These could be, for example, the type of a catalyst used in the process. This could be a blocking factor. Presently, its impact is confounded with the treatments. If we could separate its impact, we would be in a better position to identify the impact of the treatments on octane rating ,which is really our objective. Suppose there are two types of catalyst. By blocking on the catalyst level, we would choose a catalyst and then randomly assign the treatments within that setting. Through blocking, the variation between blocks is partitioned out from the experimental error.

If we do not suspect any nuisance variables in the process, meaning those beyond the factors of our study, that have an impact on the response variable, a completely randomized design should be chosen. Alternatively, if there are other factors in the process, which we may not be interested in studying but they do impact the response variable, a randomized block design could be chosen.

12-8. In a randomized block design, only one blocking variable is used, which may have several levels. However, in a Latin square design, two blocking variables are used, which is an advantage specially if one suspects that there is a second variable, not a design factor, which possibly impacts the response variable. This accomplishes more reduction in the sum of squares of the experimental error.

There are some disadvantages of the Latin square design. The number of classes or levels of each blocking variable must be equal to each other and also to the number of treatments, leaving few degrees of freedom for the experimental error. Also, there is the assumption of no interaction between the blocking variables or between each blocking variable and the treatment.

12-9. When the interaction effects are significant in a factorial experiment, it is the joint impact of the factors that impacts the response variable in a significant manner. In most processes, one cannot measure the impact of a factor, in isolation to other factors. There could be synergistic or antagonistic effects. Hence, if the interaction effects are significant, decisions on the parameter levels should be made on the basis of the interaction plots rather than the main effects.

12-10. Contrasts allow us to test a variety of hypothesis, other than the routine pair-wise comparison of treatment means. In situations involving quantitative factors, it may be of interest to determine if the mean of one treatment differs from the average of two other treatment means. Such hypothesis testing is possible using the notion of contrasts.

Orthogonal contrasts have certain properties that must be satisfied. The sum of the products of the corresponding coefficients associated with the treatment means should equal zero. Orthogonal contrasts are useful to decompose the treatment sum of squares in an analysis of variance procedure. They assist in testing hypothesis on the effectiveness of the selected treatments. With p treatments, it is possible to have (p–1) orthogonal contrasts such that the sum of squares of these contrasts equals the treatment sum of squares (SSTR). Thus:

$$SSTR = S_1{}^2 + S_2{}^2 + \cdots + S_{p-1}^2,$$

where S_i^2 represents the sum of squares of contrast i.

12-11. In a 2^k experiment, there are k factors, each at 2 levels. In a 2^{k-2} fractional factorial experiment, only $1/4^{th}$ of the total experiments (in a 2^k experiment) are selected for experimentation. Such fractionalization takes place when it is not feasible, due to cost or other resources, to run all the 2^k runs of the full factorial experiment.

Construction of fractional factorial experiments is accomplished through selection of a defining contrast or a generator. This generator may typically be a high-order interaction whose effect may not be significant and we may not be interested in. The more the fractionalization, the more the degree of confounding. Here, we will need two generators to determine the runs associated with a 2^{k-2} experiment from a 2^k experiment.

12-12. Fractionalization creates confounding. Since all treatments are not run in a fractional factorial experiment, we will not be able to estimate all of the treatment effects. In fact, some of the treatment effects will be confounded (hidden or aliased) with others. As the degree of fractionalization increases (1/2, 1/4, 1/8, $\cdots$), the degree of confounding increases too. For example, if the factor A is confounded with CD, we will not be able to distinguish between the two when making inferences. So, even if factor A may "seem" to be significant, we cannot be sure of this. It could be that either A or CD or both are significant.

12-13. A defining contrast or generator is used to decide on the treatments to select in a fractional factorial experiment from the treatments in a full factorial experiment. For example, consider a 2^3 experiment with three factors A, B, and C. For a half-replicate (2^{3-1}) of this experiment, suppose the generator chosen is the three-way interaction (I=ABC). It can be shown that for I=ABC, the selected treatments will be a, b, c, and abc. All of these treatments have a plus sign associated with the contrast ABC (principal fraction). Alternatively, the other fraction {(1), ab, ac, bc} of treatments have a minus sign associated with the contrast ABC. Hence, the generator I= –ABC, yields the other portion of the treatments called the alternate fraction.

12-14. Taguchi's concept of the loss function is based on the philosophy that any deviation of the quality characteristic from the target value causes a loss. Since noise factors cannot be eliminated, Taguchi's concept of robust design seeks to minimize loss to determine the optimal level of the design factors and also to create a design that is insensitive to the noise factors.

12-15. For the target is best situation, an example in the hospitality industry could be the temperature or humidity setting in a room. For smaller is better, an example is the cost/night charged by the establishment, while an example for larger is better could be efficiency of hotel staff/management.

12-16. Taguchi's signal to noise ratio (S/N) combines two characteristics into one measure. Signal is a measure of the average value of the desired characteristic, while noise measures the variation of characteristic. We desire the variation in the characteristic to be small. Depending on the situation, target is best, smaller is better, or larger is better, Taguchi proposes various forms of the S/N ratio and describes them as performance statistics.

In Taguchi's two-step parameter design approach, in the first step design parameter levels are selected so as to maximize the performance statistic, such as the S/N ratio. On the second step, control or adjustment parameter levels are selected to shift the average response to the target value without increasing the variability in performance.

12-17. a)   The analysis of variance table is shown in Table 12-1. For a chosen level of significance $\alpha = 0.10$, $F_{0.10,2,12} = 2.81$. Since the test statistic of $23.57 > 2.81$, we reject the null hypothesis of equality of all treatment means.

b)   95% CI for μ_1 is:

$$0.022 \pm t_{0.025,12} \sqrt{0.0042/5} = 0.022 \pm 2.179(0.009165)$$

$$= 0.022 \pm 0.01997 = (0.00203, 0.04197).$$

c)   90% CI for $(\mu_2 - \mu_3)$ is:

$$(0.108 - 0.044) \pm t_{0.05,12} \sqrt{2(0.00042)/5}$$

$$= 0.064 \pm 1.782(0.01296) = 0.064 \pm 0.0231 = (0.0409, 0.0871).$$

d)   The company that has the smallest mean degree of lateness is company 1, which should be chosen.

12-18. a)   The analysis of variance table is shown in Table 12-2. At the 5% level of significance, $F_{0.05,3,8} = 4.07$. Since the test statistic of $12.99 > 4.07$, we reject the null hypothesis of equality of all treatment means.

b)   90% CI for μ_2 is:

$$3.00 \pm t_{0.05,8} \sqrt{3/3} = 3.00 \pm 1.860 = (1.14, 4.86).$$

c)   95% CI for $(\mu_1 - \mu_2)$ is:

$$(11.667 - 3.000) \pm t_{0.025,8} \sqrt{2(3)/3}$$

$$= 8.667 \pm 2.306(1.4142) = 8.667 \pm 3.261 = (5.406, 11.928).$$

TABLE 12-1. ANOVA Table for Lateness Analysis

Source of variation	Degrees of freedom	Sum of squares	Mean square	F-statistic
Treatments	2	0.01996	0.00998	23.57
Error	12	0.00508	0.00042	
Total	14	0.02504		

TABLE 12-2. ANOVA Table for Passengers Bumped

Source of variation	Degrees of freedom	Sum of squares	Mean square	F-statistic
Treatments	3	116.917	38.972	12.99
Error	8	24.000	3.000	
Total	11	140.917		

Since the confidence interval does not include 0, there is a significant difference in these two means at the 5% level of significance.

d)   Software package 2 has the lowest average number of passengers bumped and so should be chosen.

12-19. a)   The analysis of variance table is shown in Table 12-3. At the 10% level of significance, $F_{0.10,2,4} = 4.32$. Since the test statistic of $11.81 > 4.32$, we reject the null hypothesis of equality of mean performance ratings associated with the training programs.

b)   90% CI for $(\mu_1 - \mu_3)$ is:

$$(77.333 - 87.333) \pm t_{0.05,4} \sqrt{2(10.111)/3}$$

$$= -10 \pm 2.132(2.5963) = -10 \pm 5.535 = (-15.535, -4.465).$$

Since the confidence interval does not include 0, there is a significant difference in the mean performance ratings of these two programs.

12-20. a)   The analysis of variance table is shown in Table 12-4. At the 5% level of significance, $F_{0.05,3,12} = 3.49$. Since the test statistic of $8.49 > 3.49$, we reject the null hypothesis of equality of mean reduction in blood sugar levels due to the diet types.

TABLE 12-3. ANOVA Table for Training Programs

Source of variation	Degrees of freedom	Sum of squares	Mean square	F-statistic
Training programs	2	238.889	119.444	11.81
Years of experience (blocks)	2	369.555	184.777	
Error	4	40.444	10.111	
Total	8	648.888		

TABLE 12-4. ANOVA Table for Blood Sugar Analysis

Source of variation	Degrees of freedom	Sum of squares	Mean square	F-statistic
Diet type	3	1363.75	454.583	8.49
Age group (blocks)	4	2417.50	604.375	
Error	12	642.50	53.543	
Total	19	4423.75		

b) 90% CI for μ_1 is:

$$48.00 \pm t_{0.05,12} \sqrt{2(53.542)/5}$$

$$= 48.00 \pm 1.782(3.272) = 48.00 \pm 5.831 = (42.169, 53.831).$$

c) 95% CI for $(\mu_2 - \mu_4)$ is:

$$(63.00 - 40.00) \pm t_{0.025,12} \sqrt{2(53.542)/5}$$

$$= 23 \pm 2.179(4.6278) = 23 \pm 10.084 = (12.916, 33.084).$$

Since the confidence interval does not include 0, there is a significant difference in the mean reduction in blood sugar levels between diet types 2 and 4.

d) Diet type 2 has the largest average reduction in blood sugar level and would be preferred.

12-21. a) The analysis of variance table is shown in Table 12-5. At the 5% level of significance, $F_{0.05,3,6} = 4.76$. Since the test statistic of $20.122 > 4.76$, we reject the null hypothesis of equality of mean computational times of the four software packages.

b) 95% CI for μ_C is:

$$17.85 \pm t_{0.025,6} \sqrt{16.083/4}$$

$$= 17.85 \pm 2.447(2.005) = 17.85 \pm 4.9067 = (12.9433, 22.7567).$$

c) 95% CI for $(\mu_A - \mu_C)$ is:

$$(37.425 - 17.85) \pm t_{0.025,6} \sqrt{2(16.083)/4}$$

$$= 19.575 \pm 2.447(2.8357) = 19.575 \pm 6.939 = (12.636, 26.514).$$

Since the confidence interval does not include 0, there is a significant difference in the mean computational times of software packages A and C.

d) Software package C has the smallest average computational time, a value of 17.85, with the average for packages B, D, and A being 19.475, 21.925 and 37.425, respectively. The difference in the sample averages of B and C is 1.625, of D and C is 4.075, and of A and C is 19.575. Using the information from part (c), at the 5% level of significance, there is no significant difference in the means of B and C, and D and C, while there is a significant differences in the means of A and C. Thus while software package C is preferred, based on the smallest average computational time, if B or D are attractive due to other reasons, they could be chosen. Package A would not be selected.

e) To investigate the effectiveness of the row blocking variable (problem type), we have:

$F = 13.262/16.083 = 0.824$.

At the 5% level of significance, $F_{0.05,3,6} = 4.76$. Since $0.824 < 4.76$, the use of problem type as a blocking variable has not been effective. For the column blocking variable (operating system configuration), we have:

$F = 13.356/16.083 = 0.830 < 4.76$.

So, the use of operating system configuration as a blocking variable has not been effective.

f) If no other variables are introduced, the experiment could be conducted as a completely randomized design with the software packages being the treatments. Alternatively, other effective blocking variables could be investigated.

TABLE 12-5. ANOVA Table for Computational Times

Source of variation	Degrees of freedom	Sum of squares	Mean square	F-statistic
Problem type (rows)	3	39.7869	13.262	
Operating system (columns)	3	40.0669	13.356	
Software packages (treatments)	3	970.8819	323.627	20.122
Error	6	96.4987	16.083	
Total	15	1147.2344		

12-22. a) The analysis of variance table is shown in Table 12-6. At the 5% level of significance, $F_{0.05,4,36} \cong 2.642$. In testing for interaction between temperature and pressure, $F = 47.91 > 2.642$. We conclude that the interaction effects are significant.

b) Since the interaction effects are significant as found in part (a), we take into account the joint effect of temperature and pressure. When the temperature is at $250°\,C$ and pressure is at 150 kg/cm^2, the sample average is maximum (a value of 82.6). This is the desired setting.

c) 90% of CI for μ_{13} is:

$$78.80 \pm t_{0.05,36} \sqrt{31.644/5}$$

$$= 78.80 \pm (1.6888)(2.5157) = 78.80 \pm 4.2485 = (74.5515, 83.0485).$$

d) 95% CI for ($\mu_{23} - \mu_{32}$):

$$(82.60 - 67.60) \pm t_{0.025,36} \sqrt{2(31.644)/5}$$

$$= 15.00 \pm (2.0282)(3.5577) = 15.00 \pm 7.2158 = (7.7842, 22.2158).$$

12-23. a) We have a two-factor factorial experiment using a randomized block design, where the five automobiles serve as blocks. The analysis of variance table is shown in Table 12-7. At the 5% level of significance, $F_{0.05,4,32} = 2.674$. In testing for interaction between additive and catalyst, $F = 72.11 > 2.674$. We conclude that the interaction effects are significant and so do not comment on the individual main effects.

b) 95% CI for μ_{12}:

$$60.00 \pm t_{0.025,32} \sqrt{8.7403/5}$$

$$= 60.00 \pm 2.0372(1.3221) = 60.00 \pm 2.693 = (57.307, 62.693).$$

TABLE 12-6. ANOVA Table for Ductility

Source of variation	Degrees of freedom	Sum of squares	Mean square	F-statistic
Temperature (Factor A)	2	1154.533	577.267	18.24
Pressure (Factor B)	2	4732.133	2366.067	74.77
Interaction (AB)	4	6064.133	1516.033	47.91
Error	36	1139.200	31.644	
Total	44	13090.000		

c)   95% CI for ($\mu_{13} - \mu_{23}$):

$$(45.00 - 59.20) \pm t_{0.025,32} \sqrt{2(8.7403)/5}$$

$$= -14.2 \pm 2.0372(1.8698) = -14.2 \pm 3.809 = (-18.009, -10.391).$$

Since the confidence interval does not include 0, there is a significant difference in the means.

d)   Since the interaction effects between additive and catalyst were found to be significant, we consider the treatment means to take into account the joint effect of additive and catalyst. The sample average efficiency rating is maximum, a value of 79.4, when additive is at level 3 and catalyst at level 2. This is the desirable setting.

e)   In testing the effect of blocking using 5 automobiles,

F = 34.478/8.7403 = 3.94.

At a 5% level of significance, $F_{0.05,4,32} = 2.674$. Since 3.94 > 2.674, the effect of blocking was significant. As can be observed, by blocking, the MSE has been reduced from 11.600 (in Example 12-4) to 8.7403.

12-24. a)   The contrast is L = $2\mu_3 - \mu_1 - \mu_2$. An estimate of L is:

$$\hat{L} = 2(0.044) - 0.022 - 0.108 = -0.042$$

$$Var(\hat{L}) = [2^2/5 + (-1)^2/5 + (-1)^2/5](0.00042) = 0.000504.$$

The test statistic is:

$$t = -0.042/\sqrt{0.000504} = -1.8708.$$

TABLE 12-7. ANOVA Table for Fuel Efficiency

Source of variation	Degrees of freedom	Sum of squares	Mean square	F-statistic
Additive (Factor A)	2	64.133	32.066	3.67
Catalyst (Factor B)	2	3328.133	1664.066	190.39
Interaction (AB)	4	2520.934	630.233	72.11
Automobiles	4	137.911	34.478	3.94
Error	32	279.689	8.740	
Total	44	6330.800		

At the 5% level of significance, $t_{0.025,12} = 2.179$. Since $|t| = 1.8708 < 2.179$, we do not reject H_o: $L = 0$.

b)   90% CI for $(2\mu_3 - \mu_1 - \mu_2)$:

$$-0.042 \pm t_{0.05,12} \sqrt{0.000504}$$

$$= -0.042 \pm 1.782(0.02245) = -0.042 \pm 0.040 = (-0.082, -0.002).$$

12-25. a)   The contrast is $L = \mu_1 + \mu_2 - \mu_3 - \mu_4$. An estimate of L is:

$$\hat{L} = 11.667 + 3 - 8.333 - 6.667 = -0.333.$$

$$Var(\hat{L}) = [1/3 + 1/3 + 1/3 + 1/3](3.000) = 4.00.$$

The test statistic is:

$$t = -0.333/\sqrt{4.00} = -0.167.$$

At the 10% level of significance, $t_{0.10,8} = 1.397$. Since $|t| = 0.167 < 1.397$, we do not reject H_o: $L = 0$.

b)   95% CI for $(2\mu_3 - \mu_1 - \mu_2)$:

$$(2(8.333) - 11.667 - 3) \pm t_{0.025,8} \sqrt{(6/3)(3.000)}$$

$$= 2.00 \pm (2.306)(2.449) = 2.00 \pm 5.648 = (-3.648, 7.648).$$

12-26.   The contrast is $L = 2\mu_1 - \mu_2 - \mu_3$. An estimate of L is:

$$\hat{L} = 2(77.333) - 75.666 - 87.333 = -8.333.$$

$$Var(\hat{L}) = [4/3 + 1/3 + 1/3](10.111) = 20.222.$$

90% CI for $(2\mu_1 - \mu_2 - \mu_3)$:

$$-8.333 \pm t_{0.05,4} \sqrt{20.222}$$

$$= -8.333 \pm (2.132)(4.4969) = -8.333 \pm 9.587 = (-17.920, 1.254).$$

12-27. a) The two contrasts are: $\hat{L}_1 = T_1 - T_3$, $\hat{L}_2 = T_1 + T_3 - 2T_2$. Note that $\hat{L}_1$ and $\hat{L}_2$ are two orthogonal contrasts. We have $\hat{L}_1 = 232 - 262 = -30$, $\hat{L}_2 = 232 + 262 - 2(227) = 40$. Now, $D_1 = (3 + 3) = 6$, $D_2 = (3 + 3 + 4(3)) = 18$. So, sum of squares for the two contrasts are:

$$S_1^2 = \hat{L}_1^2/D_1 = (-30)^2/6 = 150.$$

$$S_2^2 = \hat{L}_2^2/D_2 = (40)^2/18 = 88.889.$$

b) The second contrast indicates the null hypothesis H_0: $\mu_1 + \mu_3 - 2\mu_2 = 0$. An estimate of this contrast of means $= 77.333 + 87.333 - 2(75.667) = 13.333$. The variance of this contrast of means $= [1/3 + 1/3 + 4/3](10.111) = 20.222$. So, the test statistic is:

$$t = 13.333/\sqrt{20.222} = 2.965.$$

At the 5% level of significance, $t_{0.025,4} = 2.776$. Since $t = 2.965 > 2.776$, we reject H_0. Thus, there is a difference in the average ratings of training programs 1 and 3 from that of training program 2.

12-28. a) The three contrasts of totals are: $\hat{L}_1 = T_1 + T_3 - T_2 - T_4$, $\hat{L}_2 = T_1 + T_2 - T_3 - T_4$, $\hat{L}_3 = T_2 + T_3 - T_1 - T_4$. Note that these are mutually orthogonal contrasts. We have $\hat{L}_1 = 240 + 250 - 315 - 200 = -25$, $\hat{L}_2 = 240 + 315 - 250 - 200 = 105$, $\hat{L}_3 = 315 + 250 - 240 - 200 = 125$. Also, $D_1 = (5 + 5 + 5 + 5) = 20$, $D_2 = 20$, and $D_3 = 20$. The sum of squares due to each of the contrasts are:

$$S_1^2 = \hat{L}_1^2/D_1 = (-25)^2/20 = 31.25.$$

$$S_2^2 = \hat{L}_2^2/D_2 = (105)^2/20 = 551.25.$$

$$S_3^2 = \hat{L}_3^2/D_3 = (125)^2/20 = 781.25.$$

From Problem 12-20, note that the treatment sum of squares due to the diet types is 1363.75.. The above three sum of squares represent corresponding proportions of 0.0229, 0.4042, and 0.5729, respectively of this treatment sum of squares.

b) The first contrast indicates the null hypothesis H_0: $\mu_1 + \mu_3 - \mu_2 - \mu_4 = 0$. An estimate of the corresponding contrast of means $= -25/5 = -5$. The variance of this contrast of means $= [1/5 + 1/5 + 1/5 + 1/5](53.542) = 42.8336$. So, the test statistic is:

$$t = -5/\sqrt{42.8336} = -0.764.$$

At the 5% level of significance, $t_{0.025,12} = 2.179$. Since $|t| = 0.764 < 2.179$, we do not reject H_o.

c)   The second contrast indicates the null hypothesis H_o: $\mu_1 + \mu_2 - \mu_3 - \mu_4 = 0$. An estimate of the corresponding contrast of means = 105/5 = 21. The variance of this contrast of means = $[1/5 + 1/5 + 1/5 + 1/5](53.542) = 42.8336$.

So, the test statistic is:

$$t = 21/\sqrt{42.8336} = 3.209.$$

At the 10% level of significance, $t_{0.05,12} = 1.782$. Since $t = 3.209 > 1.782$, we reject H_o.

12-29. Suppose the four factors are A, B, C, and D, each at two levels. The 16 treatment combinations are: (1), a, b, ab, c, ac, bc, abc, d, ad, bd, abd, cd, acd, bcd, and abcd.

12-30. a)   We have a 2^3 factorial experiment. The main effect of A is found from:

$$A = \frac{1}{3(4)} [a + ab + ac + abc - (1) - b - c - bc]$$

$$= \frac{1}{12} [64 + 131 + 149 + 66 - 80 - 87 - 74 - 156]$$

$$= \frac{1}{12} (13) = 1.083.$$

Similarly, the other main effects and interaction effects are found and shown in Table 12-8.

TABLE 12-8. ANOVA Table for Emission Levels

Source of variation	Effect	Degrees of freedom	Sum of squares	Mean square	F-statistic
A	1.083	1	7.0417	7.0417	0.57
B	6.083	1	222.0417	222.0417	18.00
C	6.917	1	287.0417	287.0417	23.27
AB	-8.750	1	459.3750	459.3750	37.25
AC	-3.583	1	77.0417	77.0417	6.25
BC	-6.25	1	234.3750	234.3750	19.00
ABC	-18.75	1	2109.3750	2109.3750	171.03
Error		16	197.3333	12.3333	
Total		23	3593.6250		

b) Sum of squares for A is found as: $(13)^2/[3(8)] = 7.0417$. Similarly, sum of squares for all the main effects and interaction effects are found and are shown in Table 12-8. The total sum of squares is calculated as:

$$SST = [30^2 + 24^2 + 26^2 + 18^2 + \ldots + 20^2 + 22^2] - (807)^2/24 = 3593.625.$$

The error sum of squares is found by subtraction and is equal to 197.3333.

c) At the 5% of significance, $F_{0.05,1,16} = 4.4967$. Comparing the F-statistic with this value we find that the interaction effects of AB, AC, BC, and ABC are significant. The F-values associated with the main effects of B and C also exceed the critical value. However, since the interaction effects are significant, we do not make any definite inferences on the main effects.

12-31. Table 12-9 and Table 12-10 show the table of coefficients for the orthogonal contrasts.

Contrast (ABCD) = (1) – a – b + ab – c + ac + bc – abc – d + ad + bd – abd + cd – acd – bcd + abcd.

To confound the design into two blocks, we take the treatments with a positive sign in the contrast ABCD and put them in block 1, and the treatments with a negative sign are put in block 2. So the treatments in block 1 will be (1), ab, ac, bc, ad, bd, cd, and abcd. The treatments in block 2 will be a, b, c, abc, d, abd, acd,, and bcd.

12-32. Using AB as the confounding contrast, we have:

Contrast (AB) = (1) – a – b + ab + c – ac – bc + abc + d – ad – bd + abd + cd – acd – bcd + abcd.

To confound the experiment into two blocks, block 1 will contain the following treatments: (1), ab, c, abc, d, abd, cd, and abcd. Block 2 will consist of the following treatments: a, b, ac, bc, ad, bd, acd, and bcd.

TABLE 12-9. Coefficients for Orthogonal Contrasts

Contrast	Treatment							
	(1)	a	b	ab	c	ac	bc	abc
A	–	+	–	+	–	+	–	+
B	–	–	+	+	–	–	+	+
AB	+	–	–	+	+	–	–	+
C	–	–	–	–	+	+	+	+
AC	+	–	+	–	–	+	–	+
BC	+	+	–	–	–	–	+	+
D	–	–	–	–	–	–	–	–
AD	+	–	+	–	+	–	+	–
BD	+	+	+	+	–	–	–	–
CD	+	+	+	+	–	–	–	–

TABLE 12-10. Coefficients for Orthogonal Contrasts (continued)

Contrast	Treatment							
	d	ad	bd	abd	cd	acd	bcd	abcd
A	−	+	−	+	−	+	−	+
B	−	−	+	+	−	−	+	+
AB	+	−	−	+	+	−	−	+
C	−	−	−	−	+	+	+	+
AC	+	−	+	−	−	+	−	+
BC	+	+	−	−	−	−	+	+
D	+	+	+	+	+	+	+	+
AD	−	+	−	+	−	+	−	+
BD	−	−	+	+	−	−	+	+
CD	−	−	−	−	+	+	+	+

To estimate the effect of factor A, the contrast is:

Contrast (A) = a + ab + ac + abc + ad + abd + acd + abcd − (1) − b − c − bc − d − bd − cd − bcd.

12-33. Contrast (BC) = (1) + a + bc + abc + d + ad + bcd + abcd − b − ab − c − ac − bd − abd − cd − acd. So, the treatment combinations in a 2^{4-1} fractional factorial experiment using BC as a defining contrast, would be (1), a, bc, abc, d, ad, bcd, and abcd. Let us find the alias structure, where I = BC: A = ABC, B = C, D = BCD, AB = AC = ABC, AD = ABCD, BD = CD, and ABD = ACD. In order to estimate the effect of contrast BC, we would need to run an experiment with the treatments corresponding to the alternate fraction (I = −BC). Then, the information from the two experiments may be combined to estimate the effect of contrast BC.

Using a second generator of AD, the treatment combinations in a 2^{4-2} fractional factorial experiment would be (1), bc, ad, and abcd. The alias structure is as follows: I = BC = AD = ABCD; A = ABC = D = BCD; B = C = ABD = ACD; AB = AC = BD = CD.

12-34. In a 2^5 factorial experiment, contrast CDE is express as follows:

Contrast (CDE) = c + ac + bc + abc + d + ad + bd + abd + e + ae + be + abe + cde + acde + bcde + abcde − (1) − a − b − ab − cd − acd − bcd − abcd − ce − ace − bce − abce − de − ade − bde − abde.

So, the treatment combinations in a 2^{5-1} fractional factorial experiment using CDE as a generator would be: c, ac, bc, abc, d, ad, bd, abd, e, ae, be, abe, cde, acde, bcde, and abcde.

Now, in a 2^5 factorial experiment, contrast AB is expressed as follows:

Contrast (AB) = (1) + ab + c + abc + d + abd + cd + abcd + e + abe + ce + abce + de + abde + cde + abcde − a − b − ac − bc − ad − bd − acd − bcd − ae − be − ace − bce − ade − bde − acde − bcde.

218

So, using AB as a second generator, the treatment combinations in a 2^{5-2} experiment would be: c, abc, d, abd, e, abe, cde, and abcde. The alias structure is as follows:

I = CDE = AB = ABCDE; A = ACDE = B = BCDE;

C = DE = ABC = ABDE; D = CE = ABD = ABCE;

E = CD = ABE = ABCD; AC = ADE = BC = BDE;

AD = ACE = BD = BCE; ACD = AE = BCD = BE.

12-35. The contrast ABC is expressed as:

Contrast (ABC) = a + b + c + abc – (1) – ab – ac – bc. So, if I = ABC is used as the defining contrast, the treatments in the 2^{3-1} experiment are: a, b, c, and abc. Let us find the alias pattern: I = ABC; A = BC, B = AC, C = AB. We cannot estimate the interaction ABC. Also, the main effects are confounded with two-factor interactions. So the main effects cannot be estimated separately from this 2^{3-1} experiment, unless the two-factor interactions are considered to be insignificant. The estimate of the effects are as follows:

$$A + BC = \frac{1}{3(2)} [a + abc - b - c] = 1/6[64 + 66 - 87 - 74]$$

$$= 1/6 \ (-31) = -5.167.$$

$$B + AC = 1/6[b + abc - a - c] = 1/6[87 + 66 - 64 - 74]$$

$$= 1/6 \ (15) = 2.5.$$

$$C + AB = 1/6[c + abc - a - b] = 1/6[74 + 66 - 64 - 87]$$

$$= 1/6 \ (-11) = -1.833.$$

The sum of squares are computed as follows:

$SS_{A+BC} = (-31)^2/[3(4)] = 80.083.$

$SS_{B+AC} = (15)^2/12 = 18.75.$

$SS_{C+AB} = (-11)^2/12 = 10.083.$

The total sum of squares is:

TABLE 12-11. ANOVA Table for Fractional Factorial Experiment

Source of variation	Effect	Degrees of freedom	Sum of squares	Mean square	F-statistic
A + BC	−5.167	1	80.083	80.083	8.981
B + AC	2.5	1	18.750	18.750	2.103
C + AB	−1.833	1	10.083	10.083	1.131
Error		8	71.334	8.917	
Total		11	180.250		

$$SST = (18^2 + 22^2 + \ldots + 20^2 + 22^2) - (291)^2/12$$

$$= 7237 - 7056.75 = 180.25.$$

The sum of squares for the error by subtraction is SSE = 71.334. The analysis of variance table is shown in Table 12-11. At the 5% level of significance, $F_{0.05,1,8} = 5.32$. So the effect of (A + BC) is significant. The main effect of A cannot be estimated separately. In order to isolate the effect of A, four more treatments using the defining contrast as I = − ABC would need to be run. Then, of course, we have the full 2^3 experiment, for which the main effect of A was found in Problem 12-30 to be not significant. However, the effect of the interaction BC was found to be significant.

12-36. The proportionality constant in the loss function, where target is best, is:

$$k = 10/(0.0004)^2 = 62.5 \times 10^6.$$

$$E(y - 0.005)^2 = [(0.0048 - 0.005)^2 + (0.0053 - 0.005)^2 + \ldots.$$

$$+ (0.0054 - 0.005)^2 + (0.0052 - 0.005)^2]/10$$

$$= 5 \times 10^{-8}.$$

So, the average loss per reel = $62.5 \times 10^6 \times 5 \times 10^{-8} = \$3.125$.

12-37. The mean square deviation around the target value of 0.005 for the new process is estimated as follows:

$$\Sigma (y_i - 0.005)^2/8 = [(0.0051 - 0.005)^2 + (0.0048 - 0.005)^2 + \ldots$$

$$+ (0.0050 - 0.005)^2 + (0.0049 - 0.005)^2]/8$$

$$= 2 \times 10^{-8}.$$

The expected loss per reel = $62.5 \times 10^6 \times 2 \times 10^{-8} = \$1.25$. The additional cost per reel for the new process = $0.03 \times 200 = 6$. So, the new process has an increased average cost per reel of $(7.25 - 3.125) = \$4.125$, compared to the old process. The new process is not cost effective. The added expected annual costs using the new process = 10,000 x 4.125 = $41,250.

12-38. The loss function is given by:

$$L(y) = 62.5 \times 10^6 (y - 0.005)^2.$$

Let the manufacturer's tolerance be given by $0.005 \pm \delta$. We have $2 = 62.5 \times 10^6 (\delta^2)$, yielding $\delta = 0.1788 \times 10^{-3}$. So, the manufacturer's tolerance should be: $0.005 \pm 0.1788 \times 10^{-3}$.

12-39. With the process mean at the target value, the mean square deviation is estimated as the square of the standard deviation = $(0.018)^2 = 3.24 \times 10^{-4}$. The expected loss per reel = $62.5 \times 10^6 \times 3.24 \times 10^{-4} = \$202.5 \times 10^2$. So the average loss per reel for the modified process = $20250 + 1.50 = \$20251.50$. The original process had an average loss per reel of $3.125. Therefore, it would not be cost effective to make the change. The added annual loss is: $10,000 \times 20248.375 = \$20.248 \times 10^7$.

12-40. The proportionality constant for the factor of price, where smaller is better, is:

$$k_1 = 50/8^2 = 0.78125.$$

The mean square deviation for price is estimated as:

$$\Sigma y_i^2 /10 = [6.50^2 + 8.20^2 + \ldots + 7.40^2 + 8.30^2]/10$$

$$= 51.108.$$

So, the expected loss per customer due to the factor of price:

$$= 0.78125 \times 51.108 = 39.928.$$

The proportionality constant for the factor of service time is:

$$k_2 = 40/10^2 = 0.40.$$

The mean square deviation for service time is estimated as:

$$\Sigma y_i^2 /10 = [5.2^2 + 7.5^2 + \ldots + 12.0^2 + 8.5^2]/10$$

$$= 84.707.$$

So, the expected loss per customer due to the factor of service time = $0.40 \times 84.707 = 33.8828$.

Hence, the total expected loss per customer due to the factors of price and service time is $(39.928 + 33.8828) = \$73.8108$. The expected monthly loss $= 2000(73.8108) = \$147,621.60$.

12-41. The mean square deviation for service time with the added personnel is estimated as:

$$\Sigma y_i^2 / 10 = [8.4^2 + 5.6^2 + \ldots + 6.4^2 + 7.5^2]/10$$

$$= 49.396.$$

The expected loss per customer due to the factor of service time:

$$= 0.40 \times 49.396 = 19.7584.$$

The total expected loss per customer due to the factors of price and service time as well as the additional cost of personnel is:

$$(39.928 + 19.7584 + 0.50) = \$60.1864 < \$73.8108.$$

So, it is cost effective to add personnel. The total expected monthly loss $= 2000 \times 60.1864 = \$120,372.80$.

12-42. Since there is a total of seven factors, which includes the three interaction effects of importance, the $L^8(2^7)$ orthogonal array is chosen. The factor assignments for the 8 experiments are shown in Table 12-12.

12-43. There is a total of seven factors, that includes two interaction effects believed to be of importance. The $L_8(2^7)$ orthogonal array is chosen. The linear graph for this design shows the assignment of 3 interaction terms. Here, we need to estimate only 2 interaction terms. So, for the interaction that is not used, we remove the associated two column numbers and treat them as individual points assigned to the main effects. The design is shown in Table 12-13. We initially assign factor B to column 1, factor C to column 2, and factor E to column 4. Then, the interaction BxC will be assigned to column 3 and the interaction BxE will be assigned to column 5. Now, factors A and D are assigned to columns 6 and 7, respectively.

TABLE 12-12. Experimental Design for EPA

Experiment number	Factor						
	A	B	AxB	C	AxC	BxC	D
1	1	1	1	1	1	1	1
2	1	1	1	2	2	2	2
3	1	2	2	1	1	2	2
4	1	2	2	2	2	1	1
5	2	1	2	1	2	1	2
6	2	1	2	2	1	2	1
7	2	2	1	1	2	2	1
8	2	2	1	2	1	1	2

TABLE 12-13. Experimental Design for Baseball Exercise

Experiment number	Factor						
	B	C	BxC	E	BxE	A	D
1	1	1	1	1	1	1	1
2	1	1	1	2	2	2	2
3	1	2	2	1	1	2	2
4	1	2	2	2	2	1	1
5	2	1	2	1	2	1	2
6	2	1	2	2	1	2	1
7	2	2	1	1	2	2	1
8	2	2	1	2	1	1	2

12-44. Factor C has 3 degrees of freedom, while each of the other factors, A, B, D, and E has 1 degree of freedom, yielding a total of 7 degrees of freedom. The $L_8(2^7)$ orthogonal design has 7 degrees of freedom and so may be used. Combining columns 1, 2, and 3, a new column is created, that is assigned to factor C. This column has 4 unique levels. The other four factors of A, B, D, and E are assigned to the remaining columns. The final assignment is shown in Table 12-14.

12-45. There are 3 factors, A, B, and C, each at 3 levels, and the interaction BxC is important. The $L_9(3^4)$ orthogonal array is used and the assignments are shown in Table 12-15.

12-46. The experimental assignment of the factors is shown in Table 12-16.

The main effects of each factor at the three levels are found:

$$\overline{A_1} = (6.8 + 15.8 + 10.5)/3 = 11.033$$

$$\overline{A_2} = (5.2 + 17.1 + 3.4)/3 = 8.567$$

$$\overline{A_3} = (5.9 + 12.2 + 8.5)/3 = 8.867$$

TABLE 12-14. Experimental Design for Tourism Board

Experiment number	Factor				
	C	A	B	D	E
1	1	1	1	1	1
2	1	2	2	2	2
3	2	1	1	2	2
4	2	2	2	1	1
5	3	1	2	1	2
6	3	2	1	2	1
7	4	1	2	2	1
8	4	2	1	1	2

TABLE 12-15. Experimental Design for Library

Experiment number	Factor			
	B	C	BxC	A
1	1	1	1	1
2	1	2	2	2
3	1	3	3	3
4	2	1	2	3
5	2	2	3	1
6	2	3	1	2
7	3	1	3	2
8	3	2	1	3
9	3	3	2	1

$$\bar{B}_1 = (6.8 + 5.2 + 5.9)/3 = 5.967$$

$$\bar{B}_2 = (15.8 + 17.1 + 12.2)/3 = 15.033$$

$$\bar{B}_3 = (10.5 + 3.4 + 8.5)/3 = 7.467$$

$$\bar{C}_1 = (6.8 + 3.4 + 12.2)/3 = 7.467$$

$$\bar{C}_2 = (15.8 + 5.2 + 8.5)/3 = 9.833$$

$$\bar{C}_3 = (10.5 + 17.1 + 5.9)/3 = 11.167$$

$$\bar{D}_1 = (6.8 + 17.1 + 8.5)/3 = 10.800$$

$$\bar{D}_2 = (15.8 + 3.4 + 5.9)/3 = 8.367$$

$$\bar{D}_3 = (10.5 + 5.2 + 12.2)/3 = 9.300$$

TABLE 12-16. Experimental Design for Crude Oil Pumped

Experiment number	Factor				Barrels per day
	A	B	C	D	(in thousands)
1	1	1	1	1	6.8
2	1	2	2	2	15.8
3	1	3	3	3	10.5
4	2	1	2	3	5.2
5	2	2	3	1	17.1
6	2	3	1	2	3.4
7	3	1	3	2	5.9
8	3	2	1	3	12.2
9	3	3	2	1	8.5

The response variable, number of barrels per day, is of the type larger-is-better. The optimum settings of the design parameters, assuming interaction effects are not significant, are: factor A at level 1, factor B at level 2, factor C at level 3, and factor D at level 1.

12-47. The experimental design shown in Problem 12-46 is used. The main effects of each factor are found below:

$$\bar{A}_1 = (12.2 + 18.3 + 13.5)/3 = 14.667$$

$$\bar{A}_2 = (8.3 + 17.2 + 7.5)/3 = 11.000$$

$$\bar{A}_3 = (7.9 + 15.7 + 14.8)/3 = 12.800$$

$$\bar{B}_1 = (12.2 + 8.3 + 7.9)/3 = 9.467$$

$$\bar{B}_2 = (18.3 + 17.2 + 15.7)/3 = 17.067$$

$$\bar{B}_3 = (13.5 + 7.5 + 14.8)/3 = 11.933$$

$$\bar{C}_1 = (12.2 + 7.5 + 15.7)/3 = 11.800$$

$$\bar{C}_2 = (18.3 + 8.3 + 14.8)/3 = 13.800$$

$$\bar{C}_3 = (13.5 + 17.2 + 7.9)/3 = 12.867$$

$$\bar{D}_1 = (12.2 + 17.2 + 14.8)/3 = 14.733$$

$$\bar{D}_2 = (18.3 + 7.5 + 7.9)/3 = 11.233$$

$$\bar{D}_3 = (13.5 + 8.3 + 15.7)/3 = 12.500$$

The response variable is of the type larger-is-better. The optimum settings of the design parameters, assuming interaction effects are not significant, are: factor A at level 1, factor B at level 2, factor C at level 2, and factor D at level 1.

12-48. The experimental layout with the design and noise factors, and the calculated average of the response variable ($\bar{y}$) as well as the signal-to-noise ratio are shown in Table 12-17.

TABLE 12-17. Experimental Layout for Food Processing Plant

Outer			E	1	1	2	2		
Array (L₄)			F	1	2	1	2		
Inner array (L₉)			G	1	2	2	1		
Run	A	B	C	D	Response Variable			$\bar{y}$	S/N
1	1	1	1	1	18.5	21.2	20.5	19.3	19.875	24.335
2	1	2	2	2	16.8	17.3	20.9	18.5	18.375	20.044
3	1	3	3	3	21.1	21.8	20.6	19.4	20.725	26.233
4	2	1	2	3	20.2	17.7	19.8	20.8	19.625	23.265
5	2	2	3	1	16.2	21.5	21.2	21.4	20.075	17.799
6	2	3	1	2	18.3	18.5	17.8	17.2	17.950	29.809
7	3	1	3	2	20.6	21.4	16.8	19.5	19575	19.782
8	3	2	1	3	17.5	20.0	21.0	20.4	19.725	22.154
9	3	3	2	1	20.4	18.8	19.6	18.3	19.275	26.410

This is a case where target-is-best and the S/N ratio is calculated using $Z = 10 \log (\bar{y}^2 / s^2)$. We now calculate the average S/N ratio and the average response for each factor level.

Factor A – Average S/N

Level 1: (24.335 + 20.044 + 26.233)/3 = 23.537

Level 2: (23.265 + 17.799 + 29.809)/3 = 23.624

Level 3: (19.782 + 22.154 + 26.410)/3 = 22.782

Factor A – Average response

Level 1: (19.875 + 18.375 + 20.725)/3 = 19.658

Level 2: (19.625 + 20.075 + 17.950)/3 = 19.217

Level 3: (19.575 + 19.725 + 19.275)/3 = 19.525

Factor B – Average S/N

Level 1: 24.335 + 23.265 + 19.782)/3 = 22.461

Level 2: (20.044 + 17.799 + 22.154)/3 = 19.999

Level 3: (26.233 + 29.809 + 26.410)/3 = 27.484

Factor B – Average response

Level 1: (19.875 + 19.625 + 19.575)/3 = 19.692

Level 2: $(18.375 + 20.075 + 19.725)/3 = 19.392$

Level 3: $(20.725 + 17.950 + 19.275)/3 = 19.317$

Factor C – Average S/N

Level 1: $(224.335 + 29.809 + 22.154)/3 = 25.433$

Level 2: $(20.044 + 23.265 + 26.410)/3 = 23.240$

Level 3: $(26.233 + 17.799 + 19.782)/3 = 21.271$

Factor C – Average response

Level 1: $(19.875 + 17.950 + 19.725)/3 = 19.183$

Level 2: $(18.375 + 19.625 + 19.275)/3 = 19.092$

Level 3: $(20.725 + 20.075 + 19.575)/3 = 20.125$

Factor D – Average S/N

Level 1: $(24.335 + 17.799 + 26.410)/3 = 22.848$

Level 2: $(20.044 + 29.809 + 19.782)/3 = 23.212$

Level 3: $(26.233 + 23.265 + 22.154)/3 = 23.884$

Factor D – Average response

Level 1: $(19.875 + 20.075 + 19.275) = 19.742$

Level 2: $(18.375 + 17.950 + 19.575) = 18.633$

Level 3: $(20.725 + 19.625 + 19.725) = 20.025$

In maximizing the S/N ratio, the selected factor levels would be: factor A – level 2, factor B – level 3, factor C – level 1, factor D – level 3. In considering the average response, where the target value is 20, the selected factor levels would be: factor A – level 1, factor B – level 1, factor C – level 3, factor D – level 3. So, in considering both criteria, i.e., maximizing S/N ratio and meeting the target value, only factor D at level 3 satisfies both the criteria. For the other three factors, in order to decide between the two chosen levels, other considerations such as cost could be taken into account.

Now, let us investigate the possible interaction effects of BxE and CxF. The average responses are computed for the factor levels.

BxE interaction

For E at level 1: $\bar{B}_1 = (18.5 + 21.2 + 20.2 + 17.7 + 20.6 + 21.4)/6 = 19.933$

$\bar{B}_2 = (16.8 + 17.3 + 16.2 + 21.5 + 17.5 + 20.0)/6 = 18.217$

$\bar{B}_3 = (21.1 + 21.8 + 18.3 + 18.5 + 20.4 + 18.8)6 = 19.817$

For E at level 2: $\bar{B}_1 = (20.5 + 19.3 + 19.8 + 20.8 + 16.8 + 19.5)/6 = 19.450$

$\bar{B}_2 = (20.9 + 18.5 + 21.2 + 21.4 + 21.0 + 20.4)/6 = 20.567$

$\bar{B}_3 = (20.6 + 19.4 + 17.8 + 17.2 + 19.6 + 18.3)/6 = 18.817$

The slopes are not quite the same. For B_1: difference in average response from E_1 to E_2 is –0.483. For B_2: difference in average response from E_1 to E_2 is 2.35. So, we suspect that there is an interaction effect of BxE.

CxF interaction

For F at level 1: $\bar{C}_1 = (18.5 + 20.5 + 18.3 + 17.8 + 17.5 + 21.0)/6 = 18.933$

$\bar{C}_2 = (16.8 + 20.9 + 20.2 + 19.8 + 20.4 + 19.6)/6 = 19.617$

$\bar{C}_3 = (21.1 + 20.6 + 16.2 + 21.2 + 20.6 + 16.8)/6 = 19.417$

For F at level 2: $\bar{C}_1 = (21.2 + 19.3 + 18.5 + 17.2 + 20.0 + 20.4)/6 = 19.433$

$\bar{C}_2 = (17.3 + 18.5 + 17.7 + 20.8 + 18.8 + 18.3)/6 = 18.567$

$\bar{C}_3 = (21.8 + 19.4 + 21.5 + 21.4 + 21.4 + 19.5)/6 = 20.833$

The slopes are not quite the same. For C_1: difference in average response from F_1 to F_2 is 0.5. For C_2: difference in average response from F_1 to F_2 is –1.05. So, we suspect that there is an interaction effect of CxF.

12-49. The experimental layout with the design and noise factors and the calculated average of the response variable ($\bar{y}$) as well as the signal-to-noise ratio are shown in Table 12-18. The S/N ratio is calculated as $Z = 10 \log (\bar{y}^2 / s^2)$.

TABLE 12-18. Experimental Layout and Calculations

Outer array (L₈)			E	1	1	1	1	2	2	2	2		
			F	1	1	2	2	1	1	2	2		
			ExF	1	1	2	2	2	2	1	1		
			G	1	2	1	2	1	2	1	2		
			ExG	1	2	1	2	2	1	2	1		
Inner array (L₉)													
Run	A	B	C	D		Response Variable						$\overline{y}$	S/N
1	1	1	1	1	19.3	20.2	19.1	18.4	21.1	20.6	19.5	18.7	19.612	26.372
2	1	2	2	2	20.6	18.5	20.2	19.4	20.1	16.3	17.2	19.4	18.962	21.887
3	1	3	3	3	18.3	20.7	19.4	17.6	20.4	17.3	18.2	19.2	18.887	23.589
4	2	1	2	3	20.8	21.2	20.2	19.9	21.7	22.2	20.4	20.6	20.875	28.554
5	2	2	3	1	18.7	19.8	19.4	17.2	18.5	19.7	18.8	18.4	18.812	26.960
6	2	3	1	2	21.1	20.2	22.4	20.5	18.7	21.4	21.8	20.6	20.837	25.335
7	3	1	3	2	17.5	18.3	20.0	18.8	20.2	17.7	17.9	18.2	18.575	25.187
8	3	2	1	3	20.4	21.2	22.4	21.9	21.5	20.8	22.5	21.7	21.550	29.345
9	3	3	2	1	18.0	20.2	17.6	22.4	17.2	21.6	18.5	19.2	19.337	20.141

Factor A – Average S/N

Level 1: (26.372 + 21.887 + 23.589)/3 = 23.949

Level 2: (28.554 + 26.960 + 25.335)/3 = 26.950

Level 3: (25.187 + 29.345 + 20.141)/3 = 24.891

Factor A – Average response

Level 1: (19.612 + 18.962 + 18.887)/3 = 19.154

Level 2: (20.875 + 18.812 + 20.837)/3 = 20.175

Level 3: (18.575 + 21.550 + 19.337)/3 = 19.821

Factor B – Average S/N

Level 1: (26.372 + 28.554 + 25.187)/3 = 26.704

Level 2: (21.887 + 26.960 + 29.345)/3 = 26.064

Level 3: (23.589 + 25.335 + 20.141)/3 = 23.088

Factor B – Average response

Level 1: (19.612 + 20.875 + 18.575)/3 = 19.687

Level 2: $(18.962 + 18.812 + 21.550)/3 = 19.775$

Level 3: $(18.887 + 20.837 + 19.337)/3 = 19.687$

Factor C – Average S/N

Level 1: $(26.372 + 25.335 + 29.345)/3 = 27.017$

Level 2: $(21.887 + 28.554 + 20.141)/3 = 23.527$

Level 3: $(23.589 + 26.960 + 25.187)/3 = 25.245$

Factor C – Average response

Level 1: $(19.612 + 20.837 + 21.550)/3 = 20.666$

Level 2: $(18.962 + 20.875 + 19.337)/3 = 19.635$

Level 3: $(18.887 + 18.812 + 18.575)/3 = 18.758$

Factor D – Average S/N

Level 1: $(26.372 + 26.960 + 20.141)/3 = 24.491$

Level 2: $(21.887 + 25.335 + 25.187)/3 = 24.136$

Level 3: $(23.589 + 28.554 + 29.345)/3 = 27.163$

Factor D – Average response

Level 1: $(19.612 + 18.812 + 19.337)/3 = 19.254$

Level 2: $(18.962 + 20.837 + 18.575)/3 = 19.458$

Level 3: $(18.887 + 20.875 + 21.550)/3 = 20.437$

In maximizing S/N ratio, the selected factor levels would be: factor A – level 2, factor B – level 1, factor C – level 1, factor D – level 3. In considering the average response, where the target value is 20, the selected factor levels would be: factor A – level 2, factor B – level 2, factor C – level 2, and factor D – level 3. So, in considering both criteria, i.e., maximizing S/N ratio and meeting the target value, factor A at level 2 and factor D at level 3 satisfy both criteria. For the other two factors B and C, in order to decide between the two chosen levels, other considerations such as cost could be taken into account.

Now, let us investigate the interaction effects of AxE, BxF, ExF, and ExG. The average responses are computed for the factor levels.

AxE interaction

For E at level 1: $\overline{A_1} = (19.3 + 20.2 + 19.1 + 18.4 + 20.6 + 18.5 + 20.2 + 19.4 + 18.3$
$+ 20.7 + 19.4 + 17.6)/12 = 19.308.$

Similarly, $\overline{A_2} = 20.117$, $\overline{A_3} = 19.892$.

For E at level 2: $\overline{A_1} = 19.00$, $\overline{A_2} = 20.233$, $\overline{A_3} = 19.750$.

The slopes of the lines are not the same as factor E changes from level 1 to level 2. For A_2 and A_3, the slope of the lines are similar, while for A_1 the slope is different in sign. Interaction may exist between AxE and needs statistical testing.

BxF interaction

For F at level 1: $\overline{B_1} = 20.067$, $\overline{B_2} = 19.675$, $\overline{B_3} = 19.592$

For F at level 2: $\overline{B_1} = 19.308$, $\overline{B_2} = 19.875$, $\overline{B_3} = 19.783$

The slope of the lines for B_1 and B_2, as F changes from level 1 to level 2, are different in sign. Interaction may exist between BxF.

ExF interaction

The ExF interaction has been assigned to column 3 of the outer L_8 array. So one way to check for interaction is to determine the average response when ExF is at levels 1 and 2 and check for their equality.

ExF at level 1: Average response = 19.605

ExF at level 2: Average response = 19.828

It seems that an ExF interaction effect exists.

ExG interaction

ExG at level 1: Average response = 19.694

ExF at level 2: Average response = 19.739

An ExG interaction may exist.

TABLE 12-19. Accumulation Analysis for Textile Plant

Factor level	Accumulated outcomes			Percentage		
	Acceptable	Second class	Reject	Acceptable	Second class	Reject
A_1	1	2	0	33.333	66.667	0.0
A_2	1	1	1	33.333	33.333	33.333
A_3	1	1	1	33.333	33.333	33.333
B_1	0	2	1	0.0	66.667	33.333
B_2	1	1	1	33.333	33.333	33.333
B_3	2	1	0	66.667	33.333	0.0
C_1	1	2	0	33.333	66.667	0.0
C_2	2	1	0	66.667	33.333	0.0
C_3	0	1	2	0.00	33.333	66.667
D_1	1	1	1	33.333	33.333	33.333
D_2	2	0	1	66.667	0.0	33.333
D_3	0	3	0	0.0	100.0	0.0

12-50. An accumulation analysis is shown in Table 12-19. Management desires to eliminate rejects. So, the selected factor levels are: factor A – level 1, factor B – level 3, factor C – level 2, factor D – level 3.

12-51. An accumulation analysis is shown in Table 12-20. Selected factor levels are: factor A – level 1 or 2 or 3; other considerations such as cost could be used to select a level; factor B – level 3; factor C – level 2, factor D – level 2.

TABLE 12-20. Accumulation Analysis for Fabric Quality

Factor level	Accumulated outcomes		Percentage	
	Acceptable	Unacceptable	Acceptable	Unacceptable
A_1	1	2	33.333	66.667
A_2	1	2	33.333	66.667
A_3	1	2	33.333	66.667
B_1	0	3	0.0	100.0
B_2	1	2	33.333	66.667
B_3	2	1	66.667	33.333
C_1	1	2	33.333	66.667
C_2	2	1	66.667	33.333
C_3	0	3	0.0	100.0
D_1	1	2	33.333	66.667
D_2	2	1	66.667	33.333
D_3	0	3	0.0	100.0